History of Technology

HISTORY OF TECHNOLOGY

EDITORIAL BOARD

Editors
Dr Graham Hollister-Short and Dr Frank A.J.L. James
RICHST, Royal Institution, 21 Albemarle Street, London W1X 4BS, England

Dr Angus Buchanan
Centre for the History of Technology
University of Bath
Claverton Down
Bath BA2 7AY
England

Dr André Guillerme
2 rue Alfred Fouillée
Paris
France

Dr W.D. Hackmann
Museum of the History of Science
Broad Street
Oxford OX1 3AZ
England

Professor A.R. Hall, FBA
14 Ball Lane
Tackley
Oxfordshire OX5 3AG
England

Dr Alexandre Herlea
Conservatoire National des Arts et Métiers
292 rue Saint Martin
Paris 75003
France

Dr Bruce Hunt
Department of History
University of Texas
Austin
TX 78712-1163
USA

Professor Ian Inkster
School of Social Science and Policy
University of New South Wales
PO Box 1
Kensington
New South Wales 2033
Australia

Dr Alex Keller
Department of History of Science
University of Leicester
Leicester LE1 7RH
England

Professor Svante Lindqvist
Kungl. Biblioteket
102 41 Stockholm
Sweden

Dr Joseph Needham, FRS, FBA
The Needham Research Institute
East Asian History of Science Library
16 Brooklands Avenue
Cambridge CB2 2BB
England

Dr Carlo Poni
Departimento di Scienze Economiche
Università degli Studi di Bologna
Strada Maggiore 45
Bologna
Italy

Dr Norman A.F. Smith
London Centre for the History of Science and Technology
Sherfield Building
Imperial College
London SW7 2AZ
England

Dr Raffaello Vergani
Istituto di Studi Storici
Università de Padova
Via del Santo 28
35123 Padua
Italy

History of Technology

Volume Twelve, 1990

Edited by
Graham Hollister-Short and
Frank A.J.L. James

MANSELL

LONDON AND NEW YORK

First published 1990 by
Mansell Publishing Limited, *A Cassell imprint*
Villiers House, 41/47 Strand, London WC2N 5JE, England
125 East 23rd Street, Suite 300, New York 10010, USA

© Mansell Publishing Limited and the contributors, 1990

All rights reserved. No part of this publication may be reproduced or transmitted in any form or by any means, electronic or mechanical, including photocopy, recording or any information storage or retrieval system, without permission in writing from the publishers or their appointed agents.

British Library Cataloguing in Publication Data

History of technology.—12th volume (1990)
  1. Technology—History—Periodicals
    609   T15

ISBN 0-7201-2075-6
ISSN 0307-5451

Printed and bound in Great Britain by
Biddles Ltd, Guildford and King's Lynn

Contents

Editorial	vi
The Contributors	vii
Notes for Contributors	viii

KENNETH C. BARRACLOUGH

Swedish Iron and Sheffield Steel ... 1

IAN INKSTER

Intellectual Dependency and the Sources of Invention: Britain and the Australian Technological System in the Nineteenth Century ... 40

M.T. WRIGHT

Rational and Irrational Reconstruction: The London Sundial-Calendar and the Early History of Geared Mechanisms ... 65

J.V. FIELD

Some Roman and Byzantine Portable Sundials and the London Sundial-Calendar ... 103

R.T. McCUTCHEON

Modern Construction Technology in Low-income Housing Policy: The Case of Industrialized Building and the Manifold Links between Technology and Society in an Established Industry ... 136

Book Review by Frank A.J.L. James: André Guillerme, *Le Temps de l'Eau: La Cité, L'Eau et les Techniques: Nord de la France Fin IIIe–Début XIXe Siècle.* Eng. trans.: *The Age of Water: The Urban Environment in the North of France, AD 300–1800.* ... 177

Contents of Former Volumes ... 181

Editorial

It is with great pleasure that we offer the present volume after a hiatus of three years. The approach of the present editors is much like that of our predecessors. We place no (undue) restrictions on the length of the contributions that are offered to us, neither do we require what are often interminably prolonged refereeing processes. We do intend, however, that in future issues there should be notices of international conferences. In addition we plan to present essay-type retrospective reviews of books that seem to us to have had a seminal importance in the development of the history of technology.

We welcome papers on all aspects of technology from all cultures and civilizations. This policy includes topics where technology influences and/or is influenced by other aspects of a culture. For example, papers on the relations of religion, economic structure or science with technology would be considered.

The range of contributions in the present volume is well distributed chronologically and geographically, ranging in time from c. AD 500 to the second half of the twentieth century and drawing upon material relating to Australia, the Middle East and to Western Europe.

G.H.-S.   F.A.J.L.J.

The Contributors

Dr Kenneth C. Barraclough
Consultant, 19 Park Avenue, Chapeltown, Sheffield S30 4WH, England

Dr J.V. Field
Science Museum, Exhibition Road, London SW7 2DD, England

Professor Ian Inkster
School of Social Science and Policy, University of New South Wales, PO Box 1, Kensington, New South Wales 2033, Australia

Professor R.T. McCutcheon
Department of Civil Engineering, University of the Witwatersrand, PO Wits, 2050 South Africa

M.T. Wright
Science Museum, Exhibition Road, London SW7 2DD, England

Notes for Contributors

Contributions are welcome and should be sent to the editors. They are considered on the understanding that they are unpublished and are not on offer to another journal. Three copies should be submitted, typed in double spacing with a margin on A4 or American Quarto paper. Include an abstract of 150-200 words. Quotations when long should be inset; when short, in single quotation marks. Spelling should follow the *Oxford English Dictionary*, and arrangement H. Hart, *Rules for Compositors* (Oxford, many editions). Be clear and consistent.

All papers should be rigorously documented, with references to primary and secondary sources typed separately from the text in double spacing and numbered consecutively. Cite as follows for books:

> 1. Frank A.J.L. James (editor), *The Development of the Laboratory: Essays on the Place of Experiment in Industrial Civilisation*, (London, 1989), 54-5.

Subsequent references may be written:

> 3. James, *op. cit* (1), 43.

Only name the publisher for good reason. For theses, cite University Microfilm order number or at least Dissertations Abstract number. Standard works like DNB, DBB may be thus cited.

And as follows for articles:

> 13. Andrew Nahum, 'The Rotary Aero Engine', *Hist. Tech.*, 1986, 9:125-66.

Line drawings should be drawn boldly in black ink on stout white paper, feint-ruled paper or tracing paper. Photographs should be glossy prints of good contrast and well matched for tonal range. The place of an illustration should be indicated in the margin of the text where it should also be keyed in. Each illustration must be numbered and have a caption. Xerox copies may be sent when the article is first submitted for consideration.

Swedish Iron and Sheffield Steel

KENNETH C. BARRACLOUGH

The earliest steel produced in Britain was undoubtedly the product of the bloomery furnace, as evidenced by the steel chariot wheel tyres found in Anglesey.[1] These date from around 50 BC, and a metallurgical examination has shown them to be made by the forge welding of a number of small blooms. Each bloom has a more or less uniform carbon content, but there is variation between the blooms, with an overall range of 0.74% to 0.96% carbon.

With the coming of the blast furnace, the direct process was superseded by the indirect method, in Britain as elsewhere; continental finery type operations were carried out in Sussex in the sixteenth century. These relied on foreign workers, skilled in the art, and were on a relatively small scale; it is clear that they were most successful when the blast furnace metal was imported from South Wales, where the ore was haematite.[2]

The process which was to become particularly important in the development of Sheffield steelmaking, however, was the use of the cementation furnace to make blister steel. This relied on the introduction of carbon into bars of the virtually carbon-free wrought iron produced by the fining of cast iron. The mechanism was that of carburization by diffusion; the bars were packed in charcoal breeze in airtight chests, heated to a bright-red heat and held at a maximum temperature of 1100°C for several days. The carbon from the charcoal diffused into the solid bars of iron, slowly converting them into steel.[3] The first evidence for the process comes from Prague in 1574;[4] there are reports of its use in Nuremberg in 1601[5] and the method was patented in England in 1613 and 1617.[6] Thereafter, it was slowly adopted by the English steelmakers and there is a description of its use in the valley of the Severn in 1686.[7] From here it was taken to Birmingham and to Sheffield, but the major growth was in the Newcastle area, particularly in the valley of the Derwent, a southern tributary of the Tyne. It was in this area that the main steelmaking effort was made in the period 1700–1780; here, now, stands the sole surviving complete British cementation furnace, situated at Derwentcote (Figure 1). This furnace, seen, described and sketched by Angerstein in 1754,[8] was built around 1740, can be taken as typical of all the cementation furnaces which operated in Britain over the next 200 years; the only modification was one of scale. Angerstein described the furnace as converting ten tons of iron per heat; it must have been modified internally sometime later. Looking at it today, it has larger chests, giving an overall

Figure 1 Derwentcote furnace. This furnace was built in about 1740 on the banks of the River Derwent, a southern tributary of the Tyne, some thirteen miles from Newcastle. It is the sole surviving complete cementation furnace in Britain. It was capable of converting ten tons of bar iron (five tons in each chest) per heat and run continuously would produce up to 150 tons of blister steel per annum. It had ceased operations by 1896 and is currently being excavated and restored by English Heritage.

capacity of around fifteen tons. In addition, it has external buttresses, not shown by Angerstein, but otherwise all the features in the old sketch are recognizable. At the time Derwentcote was built, the furnaces in Sheffield were much smaller, having only half the capacity, but within the next 120 years several Sheffield furnaces were built to convert up to forty tons per heat.

The Derwentcote furnace shows an internal construction of two chests with a central firegrate and with numerous flues around the chests. The reverberatory arch above the chests has flues leading into the conical chimney above and the whole structure is designed to give reasonably uniform heating to any material contained in the chests. The chests themselves are made from slabs of fine-grained sandstone; alternatively they could be constructed from refractory bricks. The method of operation has been described fully elsewhere.[3, 11]

Whilst the production of blister steel by cementation (Figure 2) came to be

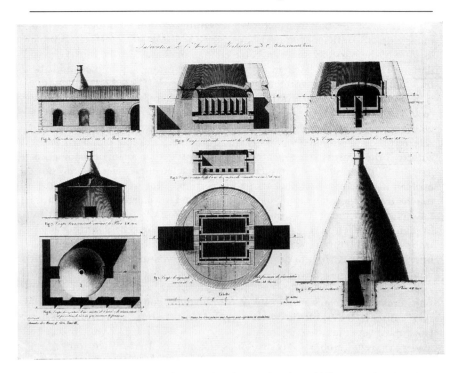

Figure 2 The cementation furnace for the production of blister steel. This plate is derived from the report by Professor Le Play on Steelmaking in Yorkshire (*Annales des Mines*, 1843) and the various diagrams clearly illustrate the internal structure of the type of furnace which had proliferated in the Sheffield area. The iron bars, packed in charcoal breeze in the two chests and then sealed to exclude air, could be raised to bright-red heat by the action of the burning coal on the firebars and the multitude of flues around the chests, the necessary draught being provided by the tall conical chimney.

looked upon as particularly English (especially after the introduction of the crucible process, which needed blister steel as its raw material) it was practised extensively elsewhere. Indeed, there was a cementation furnace operating at Davidshyttan in Dalecarlia as early as 1654; by 1764 there were twenty-one furnaces in Sweden, and ninety-three by 1860. However, they seem to have been quite small-scale undertakings since only 6,970 tons of Swedish blister steel were made in 1861,[9] whereas in 1862 the 205 furnaces in the Sheffield area produced some 78,000 tons.[10] It should be noted, however, that there are still two complete cementation furnaces at Osterby; these were wood fired and there is a good description of the operations written by the last of the Osterby steelmakers.[11]

Blister steel as it came from the furnace was an unattractive material; it was brittle, it generally had a considerably higher carbon content on the outside than in the centre and the surface was covered with blisters, hence its name. The blisters arose from the reaction of the slaggy streaks just below the surface of the bar iron with the incoming carbon, producing carbon monoxide at sufficient internal pressure to distort the surface. Simple forging of the bars gave a remarkable change in character; the blisters were flattened and welded back into the body, the grain was refined and the material became dense and tenacious. The more usual method of handling it, however, was to break off short lengths, pack them together into a bundle (referred to as a faggot) to heat this in a forge fire and then to forge weld the whole bundle together to produce what was known as shear steel (sometimes referred to as German steel). This was quite a remarkable material, particularly if the process was repeated by forging together a faggot made up of lengths of shear steel to produce double shear steel. In view of the non-uniformity of carbon content across the section of the original blister steel, the shear steel had alternating bands of high- and low-carbon material across its section and the forging of this material into knife blades, followed by hardening and tempering, gave a combination of an excellent cutting edge, from the high-carbon areas, with a high degree of ductility from the interleaved low-carbon areas. Shear steel cutlery thus gained a very high reputation.

For some applications, however, shear steel was just not satisfactory; for fine items such as clock springs its non-uniformity, coupled with the residual slag content, led to failures. It was for this reason that a Doncaster clockmaker, Benjamin Huntsman, carried out experiments which were to lead to a revolutionary advance in steelmaking. In 1740 he moved to Handsworth, a village two miles from the town of Sheffield, and attempted the melting of blister steel in a crucible, following the methods used by the brassfounders. His major problem was maintaining a sufficiently high temperature for long periods. In this he was fortunate that Abraham Darby had recently established the use of coke as a metallurgical fuel and that the glassmakers had introduced crucibles made from the very refractory Stourbridge clay. The development took him over ten years of work, but eventually he was able to set up the first crucible steelworks at Attercliffe, on the outskirts of Sheffield, in 1751. We have no record of his furnace arrangement. In 1767, however, Bengt Quist Andersson visited Sheffield and two years later set up his own crucible melting shop at Ersta; by good fortune, a drawing of this has survived (Figure 3).[12] This shows all the features which were to be observed in all coke-fired crucible furnaces for almost the next two centuries. The melting holes themselves were set into the floor of the shop, with their furnace bars almost a metre below and accessible from the cellar, which provided the air for combustion of the coke. The draught for combustion was provided by the tall chimney; at Ersta the top of the chimney was almost ten metres above ground level. It was also common practice, again as shown in the Ersta drawing, for the one chimney stack to carry the flues from several furnaces, so that the recognizable feature of crucible shops from the exterior was the rectangular chimney block, in the same way that the cementation furnaces could be recognized by their conical chimneys. As the crucible method became established in Sheffield and the number of furnaces increased it became necessary for the larger works to have their own cementation furnaces; for this reason, what might be termed

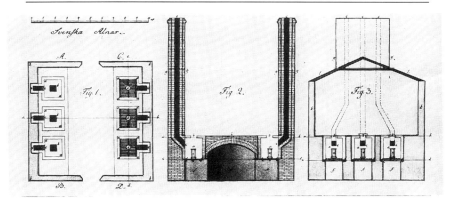

Figure 3 Crucible steel furnace established at Ersta, Stockholm, in 1769. This drawing, reproduced from Sahlin's paper on Swedish crucible steel (*Med Hammare och Fackla*, 1932) shows the furnace erected by Bengt Quist Andersson on his return from a visit to Sheffield. It is the only surviving evidence for the type of construction employed by Benjamin Huntsman at his first works in Attercliffe. The chimneys are some 27 ft tall, the crucible holes 30 in square and 36 in deep; the crucibles were 14 in high and of 7 in diameter and contained 10-12 lb metal.

the 'integrated steelworks' of the first half of the nineteenth century all had collections of both conical and rectangular chimneys and several engravings exist showing this combination. It should be remarked that, in 1862, when the 205 cementation furnaces mentioned above were producing their 78,000 tons of blister steel during the year, over 50,000 tons of this were being converted into crucible steel ingots using the 2,495 melting holes reported to be available in Sheffield at the time.[13]

The crucible melting process has been fully described elsewhere.[14] In essence, however, preheated crucibles with their lids would be placed on the furnace bars, some burning coals placed around their base and the hole then filled up with coke (Figure 4). With the cover pulled down over the hole, the full air draught allowed through the hole for a period of about thirty minutes would raise the temperature of the crucible to a white heat. The furnace cover would then be raised, the crucible lid removed and the weighed charge of blister steel pieces put into the crucible through the charging funnel. With the lid back on the charged crucible, and the space around the crucible recharged with coke, the full draught would again be applied. The coke would burn down in around one hour, and after two further charges of coke and two more hours of heating, the metal would be molten; but it required more coke and a further forty minutes to one hour at the highest possible temperature to render it ready for pouring into the ingot mould. The crucible was then pulled up out of the hole; the 'puller out'—the man performing this task—wrapped his legs with oilskin, which he covered with sacking tied on top, and tied a piece of sacking round his waist as an apron

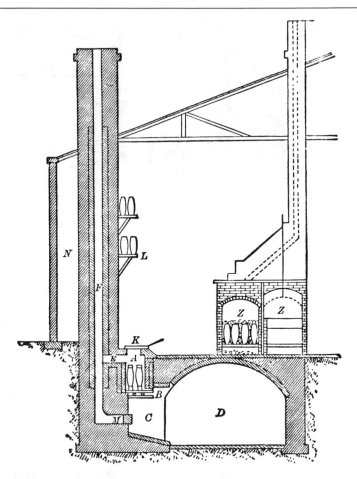

Figure 4 Coke-fired crucible furnace in Sheffield in 1870. This shows the typical fully developed form of furnace, the crucibles now containing up to 60 lb of metal and there being two per melting hole, these being oval in cross section. The furnaces were built side by side, with up to eight in a row, the flues being brought out into rectangular chimney blocks (as can be seen on a smaller scale in Figure 3). A is the crucible melting hole, lined with ganister; B the fire bars at the base of the hole; C the ashpit leading to D, the cellar, which provided the necessary air for the combustion of the coke around the crucible in the hole, the air being pulled through the flue E by the chimney F, which in this case would be anything up to 30 feet high. When the removable furnace cover, K, was taken off for operations within the hole itself, such as charging the metal to the crucible or testing the contents to determine they were completely molten, the draught through the furnace could be by-passed by removing the stopper from the flue M, the air then going straight up the chimney from the cellar. L shows the shelves over the furnace on which the moulded crucibles, after preliminary drying from the moulding operation, probably in the warm space behind the chimney, N, would be placed to dry out completely, prior to being charged, the night before they were needed for service, in the coke-fired annealing furnace Z.

and two other pieces on his arms as sleeves. All this sacking was then saturated with water prior to the cover being taken away from the furnace hole. Then, standing astride the hole, with a pair of tongs he would grip the white hot crucible and, with one lift, place it on the floor beside the crucible hole. It has to be remembered that, in the operation as carried on around 1850, the weight of charge had risen to some 25 kg (from the original 4 kg which Huntsman had melted). This meant that the crucible would weigh almost 15 kg and the tongs only a little less, so that almost 50 kg had to be lifted through a metre through a furnace atmosphere with a temperature of around 1600°C. The crucible and its contents were then picked up with a different pair of tongs by the 'teemer', whose task was to pour the steel carefully into the waiting ingot mould, without any metal catching the sides of the mould in its travel to the bottom. These ingots were then carefully forged to give the highest quality steel procurable anywhere in the world.

The combination of the cementation process with crucible melting was very much a Sheffield tradition: it is referred to in the literature as 'the Sheffield methods' and constituted *les procédés anglaises* as far as the French were concerned. It was first the cementation process and, subsequently, its combination with the Huntsman crucible process that called for the provision of a high-quality iron supply and it is in this context that the interaction of Swedish iron and Sheffield steel arose.

THE USE OF SWEDISH IRON

Seventeenth and Eighteenth Centuries

In 1631 it was reported that certain aldermen of the City of London, at the request of the Privy Council, were investigating the production of steel from Swedish iron, although no specific Swedish mark was quoted.[15]

There is evidence of a specific use of Walloon iron, however, in 1686;[16] in the description of John Heydon's making of blister steel at Bromley in the parish of Kingswinford, it is stated that he is using 'not English, but Spanish or Swedish barrs, here called bullet iron'. This, without doubt, is a reference to the iron from Osterby whose stamp mark was referred to as 'double bullet' in Sheffield steelmaking circles early in the eighteenth century and which also seems to have been known in France as *le fer des doubles boulets*.

Kingswinford is just north of Stourbridge, which became important in the second half of the seventeenth century as a centre of the infant steel industry, presumably drawing its supplies of Swedish iron from the inland port of Bewdley, on the River Severn, a few miles away. It was from Stourbridge that the steelmaker Ambrose Crowley moved at the end of the seventeenth century to commence operations some six miles from Newcastle upon Tyne at Winlaton, on the River Derwent, itself a tributary of the Tyne. In the Winlaton Council Instructions[17] there are specific references to the use of 'the best Orgroond iron, all raw ends cutt off, all flawed or cracky parts layd by or cutt off.' In addition, there appear representations of the stamp marks of both Leufsta and Gimo iron as examples (Figure 5).

From the Sheffield area, there are records of a 'Steele Trade' carried out by the John Fell partnerships; these cover the period 1699 to 1765.[18] The

Figure 5 Bar iron from the Walloon forge at Leufsta. This was the stamp mark of the most prized of all the Walloon irons coming from the Dannemora region. It was known to the Sheffield steelmakers as 'Hoop L Iron' and was always the most expensive of the Swedish bar irons.

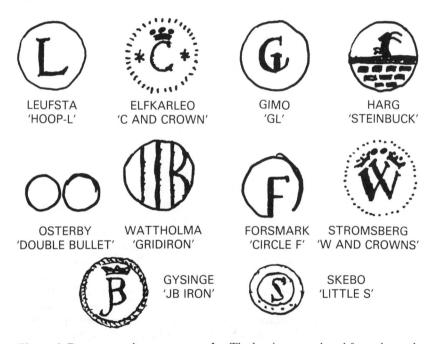

Figure 6 Dannemora iron stamp marks. The bar irons produced from charcoal-smelted cast iron from the ore mined in the Dannemora region, using the old-established Walloon refining process, were the most sought after for steelmaking on account of their freedom from sulphur and phosphorus. There were numerous forges, all of which were closely controlled as to their output so as to maintain quality, and each forge had its own distinctive stamp mark. These all became well known in Sheffield by descriptions of their marks, as can be seen here.

earlier account books generally refer to the use of 'steele iron' (although there is an isolated mention of 'Hoop L' in 1701) or slightly more definite references to 'Danks iron'—presumably iron shipped through the port of Dantzig—or 'Swedes iron'. From 1710, however, specific entries in the accounts display the stamp marks (or the names by which they became known in Sheffield for the next two centuries) for individual deliveries of iron. Over the next few years there had appeared, in addition to 'Hoop L' (Leufsta), those of 'Double Bullet' (Osterby), 'C and Crown' (Alfkarleo), 'PL' (Akerby), 'W and Crowns' (Stromsberg), 'GL' (Gimo), 'Steinbuck' (Harg) and 'Gridiron' (Wattholma) (Figures 6 and 7). This list is significant in two respects. In the first place, it is virtually a full catalogue of the Walloon iron forges operating on the Dannemora ores—the only notable exception, indeed, is that of 'Circle F' from Forsmark! Furthermore, throughout the whole of these accounts there is no single instance of any

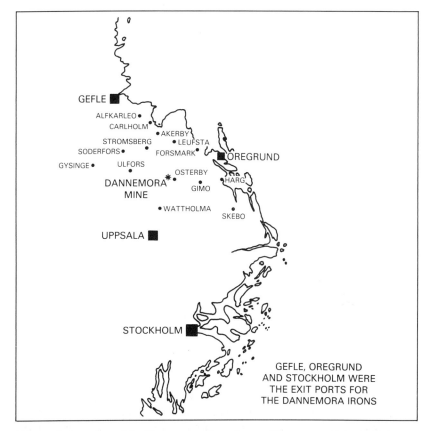

Figure 7 Location of the Dannemora iron forges. These were all sited within easy reach of the mine at Dannemora. The bar iron was originally exported from the port of Oregrund – this explains why these materials became known as the Oregrund or 'Orgroond' irons. Later considerable quantities were shipped out through Gefle (or Gavle); eventually the main trade went through the ironyards at Stockholm.

other iron being used for steelmaking; this becomes even more significant when it is realized that this 'Steele Trade' was but a minor part of the activities of the partnerships; their main operations were the smelting of the local coal measure iron ores in charcoal-fired blast furnaces and the conversion of the pig iron into wrought iron by the finery-chafery processes and, surprisingly, not one trial of their own iron seems to have been made.

Henrik Kalmeter was in England between 1719 and 1720 and again between 1723 and 1725, and left reports which confirm this universal use of Swedish iron in the production of blister steel. On his first journey he visited Blackhall Mill, a steelworks belonging to Dan Heyford some thirteen miles from Newcastle and only seven miles from Winlaton, and reported that with the rise in price of Swedish iron the cost of steel had also risen.[19] In 1725 he gave a more general account of steelmaking in England. He stated quite clearly that no English iron was converted into steel because it did not have the required strength; he also found that no attempt had been made to use Russian iron; all steel was made of Swedish or Spanish iron. The best kind of iron was tough and hard and Oregrund iron was considered to be the best of this class, with the Leufsta and Osterby marks being the most sought after; whilst the other Oregrund marks were known and good steel made from them, they were not considered as good as those two, which were the only ones used to make 'German Steel' (shear steel). The second class of iron was soft and tough but did not possess the strength of the Oregrund irons; the Spanish iron was in this class.[20]

Samuel Schroderstierna was another Swedish visitor interested in steelmaking; his journal, covering the period 1748 to 1751, reports that the steel made in Birmingham was produced from Swedish iron. English iron had been tried but found useless for steelmaking since it did not have enough of what the English termed 'body'. The favoured stamps were those from Leufsta and Akerby, the latter being considered equal to or even better than the former. The steel used at the Town Mill in Birmingham employed iron from Leufsta, Osterby and, rather surprisingly, 'PS' (Nyfors).[21] The situation in Birmingham had been discussed in Parliament in 1737, when it was stated that 220 tons of Swedish iron were converted into steel in that town during the year. Two kinds of Oregrund iron were imported:

> 'the first is generally made into steel and is the fittest for it of all the irons yet discovered and sold at Bewdley at £17.10 sometimes £18 per ton; the second which has not enough body to make steel sold at £14.10 per ton.'[22]

No specific marks are quoted. The same report also stated that the total conversion of bar iron into steel in the whole of Britain was about 1,000 tons per annum. From what is known of the furnaces in and around Sheffield in 1737, the annual production of blister steel in that region can only have been of the order 150 tons, and activities in Bristol and in London seem to have been on an even smaller scale. The major steelmaking centre in the North East, in the valley of the Derwent, can be seen to have produced some 500–600 tons per annum at that period.

The year 1740 marks the start of the revolution in steelmaking, since it was then that the clockmaker, Benjamin Huntsman, moved from Doncaster to Handsworth, a village two miles from Sheffield, and commenced his experiments which were to lead to the establishment of the crucible process. Taking the only steel available, namely blister steel, he eventually succeeded in melting it and casting the metal into ingots, setting up his first works in 1751 at Attercliffe, even nearer to the town of Sheffield. This in no way immediately altered the requirements for high-quality iron, and it should be noted that his son, William Huntsman, who carried on the business in Attercliffe, in discussion with Gustav Broling in 1798[23] attributed the renown of the Huntsman product to the exclusive use of the *three* Swedish stamps, which as a result of trials of irons from all over Europe had been proved the best. In a later passage, Broling indicates that the Huntsman concern for more than thirty years had used no other iron than the *four* stamp marks from Osterby, Gimo, Leufsta and Akerby. Carrying on the tradition, as will be detailed later, it could be noted here that the last records extant from Benjamin Huntsman and Company, dating from 1924 to 1933, still evidence this loyalty to the Swedish Walloon irons.

The travel journals of Reinhold Angerstein, who was in England between 1753 and 1755, confirm the exclusive use of Oregrund iron in steelmaking in Birmingham as well as in the North East;[24] by this time the Crowley steelmaking operations had been transferred to Swalwell, the largest steelmaking establishment in the country, which used 400 tons of Oregrund iron per annum.

During the period 1759 to 1772 the Company of Cutlers in Hallamshire operated an enterprise aimed at providing their members with blister steel at a favourable cost, and the ledgers of these operations survive.[25] From the point of view of the raw materials they are instructive. The iron used was virtually Swedish throughout, partly obtained from Samuel Wordsworth of London but mainly through Joseph Sykes of Hull. The main purchases were of Leufsta and Osterby, but there were quantities of two other cheaper grades: AOK and CDG. It seems that AOK came from Gysinge, and was an earlier stamp mark for what later came to be known as JB; CDG iron came from Ullfors which later was marketed with the W and Crown stamp mark. It is of some interest that there is evidence in these records of a slight deviation from the rule that only Swedish iron should be used for steelmaking; in each of the years 1765 and 1766 parcels of around 1.2 tons each of English iron were used as trials but no comments were made as to the results.

Johan Ludvig Robsahm visited Swalwell in 1761 and pointed out quite clearly a preference for the iron from Leufsta and Akerby for steelmaking; it is noteworthy that the lower grades of Oregrund iron were used for non-steelmaking purposes, such as Harg for anchor making (where, it is pointed out, that they dare not rely on Russian, Spanish or American iron in case it be 'cold-short') and Alfkarleo iron for the production of mattocks and hoes for America. He also reports that the last charge in the nearby cementation furnace at Blackhall Mill, now owned by a Mr Hodgson, was entirely Osterby iron, since not a single bar of either Leufsta or Akerby iron was available in either Hull or London. This preference of both Crowley and Hodgson for Akerby iron over that from Osterby is interesting; Hodgson also had a good opinion of 'W and Crown' from Vessland and wished he could obtain more of it. At a later meeting with Joseph Sykes of Hull, Robsahm learned that the entire production of the most important stamps of Oregrund iron was to be divided between merchants in Hull, London and Bristol; he expressed the opinion that this good iron should be converted to steel in Sweden—something which had to wait over a hundred years for Baron Tamm to put in hand at Osterby.[26] Robsahm also visited Sheffield, where he saw a small cementation furnace in which bars of AOK iron were to be converted. He noted that this mark of iron was said previously to have been good, but new supplies, with 'VII' stamped at the centre of the bars, were not considered to be the most suitable kind.

The next commentary on the use of iron at Swalwell comes from the Frenchman, Gabriel Jars, who was in England in 1765.[27] He reported that the one and only iron which had been found fit for conversion into steel was Swedish iron. Several trials had been made with iron produced in England but never yet had there been produced a steel of sufficiently good quality. He also pointed out that different Swedish irons were used; according to their different quality they gave varying costs of steel, since they themselves had different value. He later visited Sheffield, indicating that the furnaces there were smaller than those at Swalwell but that the method was the same, with exclusive use of Swedish iron.

Bengt Quist Andersson in his report on his visit to England in 1766 to 1767 made extensive comments on the iron used for steelmaking.[28] He stated that the selection of the iron received the closest attention, not only with regard to the quality of steel produced but also with due consideration to the ease or difficulty with which the iron might be converted into steel. If an iron could be converted to satisfactory steel but did not do so consistently, so that parts of one charge or even parts of the same bar were converted into good steel whereas the other parts were not completely converted, then the iron was considered useless as a raw material for steel. Again, if the iron required several days longer for its cementation than other tried kinds, it was rejected even though it produced good steel. Although practically all known makes of iron had been investigated in England in this way, nobody had so far succeeded in making blister steel with a combination of all the desired properties from any kind of iron other than the Walloon iron made from Dannemora in Sweden. With Biscayan iron barely half of the five tons of bars, after cementation lasting several days longer than usual, had become steel, and this was soft and brittle. Six tons of Siberian iron, after being subjected to the same trial, turned out even worse. The part of it which finally became steel consisted of open grains like spelter. Experiments on the use of both Russian and English iron for the making of blister steel had been carried out several times in Newcastle: in spite of the longer time taken before these kinds could be converted to steel, it was generally found that one end of a bar often consisted of little more than iron whilst the other end was steel. In addition, the steel product always lacked 'body'. Andersson also reported that at a steel furnace at Newcastle he had seen a variety of American iron that was to be tried for steelmaking, but it seemed to be of poor quality, full of cracks and slivers and altogether badly forged. Swedish iron generally had such a good reputation in England that blacksmiths there would not undertake certain difficult work if they could not select Swedish iron for the purpose. It was also stated that the Swedish iron was superior as regards its specific gravity. For certain purposes, nevertheless, Andersson found that Spanish iron was regarded with considerable confidence, although not for steelmaking.

Andersson further recounted the characteristics which guided the English steelmakers in their selection of good iron for conversion and the properties they considered it should have:

1. It must have 'close grains', that is, a clean dense-grained fracture, free from iron fibres. Consequently iron that is fibrous ('tough iron') is not in demand, because it requires a longer time for conversion to steel than other kinds and it can hardly ever become uniformly hard steel.
2. The finer and more uniform the grain is, as shown by the fracture, the better the iron is considered to be; conversely, an iron consisting of coarse grains ('open grains') is supposed to be of inferior quality.
3. The iron that can be converted in the shortest time is held to be the best, but is even better if . . .
4. It comes from the furnace covered by many small and closely spaced blisters.
5. Some steelmakers value more highly a steel which has a coarse fracture when it leaves the furnace, before it has been drawn down, than a steel with a finer fracture. Steelworkers prefer steel free from porosities, such as are generally indicated by the presence of large blisters on the surface.

As regards specific stamp marks, Andersson reported that in the North East those most sought after were Leufsta, Akerby and Stromsberg; these were also of good reputation in Sheffield and Birmingham, as well as Osterby, Gimo and Ranas. He also made the comment that AOK and CDG no longer retained their former high standing, the steelmakers complaining that 'their substance was burnt away', referring to iron showing fibres in the structure, which they did not like.

It is interesting to compare the above comments with those of Hassenfratz as set out in his textbook of 1812.[29] He classified all available irons into five categories:

1. Tough and soft
2. Tough and hard
3. Cold short
4. Hot short
5. Brittle, being both hot short and cold short

This classification was based on the reaction of the bars to bending to and fro, both cold and after being taken to a red heat. Only the first two categories were considered suitable for cementation, the second being considered better than the first since it took a shorter time for adequate carburization. Such categorizations can, with modern knowledge, be explained on the basis of chemical analysis. Thus the tough and soft materials would be low in carbon and contain only small quantities of the two unwanted impurities as far as the steelmaker was concerned—sulphur and phosphorus; it would seem that the Spanish iron was of this type. The tough and hard material would be similar but would contain an appreciable amount of carbon, possibly up to 0.25%; the Swedish Oregrund iron, produced by the smelting of the low-phosphorus Dannemora ores in charcoal-fired blast furnaces, followed by the refining of the pig iron by the Walloon process, was of this type. Cold short iron (material which was brittle to work when cold) generally contained an appreciable amount of phosphorus. A typical cold short iron was the kind which was produced in South Yorkshire by the Fell partnerships from the coal measure ironstones and contained 0.2–0.3% phosphorus. For the production of nails, incidentally, this was much more favoured than the Swedish iron but, as noted above, it was considered most unsuitable for steelmaking. Hot short iron (material which proved difficult to forge when hot) was characterized by a relatively high sulphur content, typical of material smelted using coke and refined in coal-fired furnaces. The early Coalbrookdale iron produced by Abraham Darby must have been of this type, since it is noted that it took several years before a suitable forge iron was produced from the coke-fired furnace—presumably by the inclusion of major quantities of lime as flux. Brittle iron—both hot short and cold short—was high in both sulphur and phosphorus and was virtually valueless; it would, it is felt, have been better to use the blast furnace product for the making of castings, rather than trouble to refine it to bar iron.

The unique value of the Swedish Walloon iron to the English maker of blister steel was, nevertheless, rather more than its freedom from sulphur and phosphorus and its somewhat elevated carbon content. Whilst its purity was essential and its higher strength, a function of the higher carbon content, facilitated the subsequent carburization in the cementation chests, there are continued references to the somewhat indefinable characteristic of 'body' in the Oregrund irons. This seems to have connections with a fine-grained structure, a well-disseminated pattern of fine slag streaks (which would subsequently show up as evenly distributed small blisters on the surface of the cemented bar) and a resistance in the blister bar to subsequent decarburization when reheated for forging operations. It may well be significant that the Dannemora ores generally contain significant proportions of manganese oxide, which is to a large degree reduced on smelting thus producing a manganiferous pig iron. The Walloon method of refining reoxidized the bulk of the manganese, producing a slag high in manganese oxide. Such a slag has two important characteristics: firstly it is less strongly oxidizing than the corresponding iron oxide slag free from manganese oxide, and thus tends to leave more carbon in the metal; in addition, and probably more importantly, this slag is more fluid and so, on forging the bloom, the slag is more completely expelled and any remaining slag is likely to be more finely disseminated. Certainly, recent work on surviving samples of such irons tends to confirm the lower slag content and the finer distribution as compared with normal English wrought irons of similar period.

The First Half of the Nineteenth Century

The surprising amount of information which exists for eighteenth-century operations ceases abruptly in about 1770. There follows a 'dark age' of some sixty years, the only information available during this period being an estimate of the amount of trade in iron bars into Sheffield likely to be carried by the proposed extension of the canal into the centre of the town. This was in 1802; the actual extension was not completed until 1819. This 'Dunn Survey'[30] highlights the users of iron and they are, in the main, the known steelmakers. The overall total is given as 3,050 tons per annum; it is reasonable to assume that this would virtually all be Walloon iron destined for steelmaking, since only 100 tons is indicated as being transported outside the Sheffield area.

It should be made quite clear that, during the period under survey, the overall picture can only be glimpsed from a relatively few chance survivals of records, and that much subjective judgment must be used in putting together a survey of the continued interrelation between Sweden and Sheffield during the remainder of the century. When the evidence, scanty as it is, begins to be available again, it is found that the situation had changed considerably. It was some years after the comments made by Andersson and only just prior to the statements of Hassenfratz that the true role of carbon in steel had become fully understood, following the work of Bergmann and Rinman in Sweden and of Berthollet and others in France. By the 1830s, the steelmaker had a much clearer idea of what he was trying to achieve. In addition, the small-scale operations in the Sheffield area had expanded considerably and now completely overshadowed those in the North-East. The underlying reasons are fairly clear. South Yorkshire had the advantages of cheap and abundant supplies of coal, there were adequate local supplies of good refractory materials, water power was plentiful and the well-wooded areas were used to provide any charcoal requirements. Only the iron supplies were missing. With the gradual improvements in communications, particularly the completion of the canal from Hull in 1819, with improvements in roads and, eventually, with the coming of the railway in 1838, Swedish iron could arrive in Sheffield without hindrance. There was, however, a further factor in that the crucible process had become firmly established as the Sheffield method and crucible steel had earned a worldwide reputation for quality. The stage was set, therefore, for Sheffield to become the steelmaking centre of Britain and, indeed, for the next quarter century, of the world.

It is worth pausing to consider the role of Swedish iron in these changed circumstances. No argument can be made against the essential requirement for a steelmaking iron with low sulphur and phosphorus contents. This was all converted into blister steel, some of which would be forged or rolled direct, but an increasing proportion was to be melted in crucibles and cast into ingots for forging. For direct use—for the making of shear steel for fine cutlery blades and so on—the original preference for Walloon iron was still valid. But what of the material used as melting stock? Looking at the situation through the eyes of a modern steelmaker, it would seem that the higher carbon content of the Walloon iron was only useful in that it reduced the conversion time in the cementation furnace. The quantity of entrained slag and its distribution pattern within the bars could be considered as relatively unimportant, since the melting operation would release it from the metal and it could be prevented from entering the ingot. Likewise the physical condition of the bar—its freedom from cracks, laps, seams and so on—provided its chemistry was right, was not, it would seem, particularly important; indeed, from the earliest days of Huntsman's own experiments onwards, the cracked and flawed ends, cut off from the blister steel bars when they were removed from the chests, provided a useful source of raw material for the crucibles at a reduced price—there are numerous references to stocks of blister bar ends at the crucible steelworks in Sheffield well into the nineteenth century.

This type of thinking obviously exercised the minds of the Sheffield steelmakers at the time when the demand for steel was growing and was no longer confined to the needs of the cutlers and edge tool makers; there were now more general applications for steel in the expanding engineering trades and particularly in the supply of material for the railways—buffers, springs and moving parts within locomotives. Here the requirement really was for a stronger and more reliable material than wrought iron, but at a reasonable price. Since the major item in the cost of the steel was the price paid for the iron used, then some economy could be made by utilizing a cheaper iron. It is not surprising, therefore, that investigation was made of the 'common' grades of Swedish iron as raw materials for steelmaking, particularly after the introduction of the Swedish Lancashire process with its improvement in this type of bar iron. As early as 1830 Daniel Doncaster was selling blister steel made with bar iron from Backefors and Svana, together with the Russian CCND, at about 60% of the price of the top-grade Dannemora blister bar.[31] The interest shown in the wider range of irons is made clear from an 1832 list of prices per ton of bar iron:[32]

Leufsta, Osterby and Gimo	£39–£40
Forsmark, Vattholma and Stromsberg	£32
Harg and Skebo	£26
Soderfors	£25
Backefors	£24
Russian CCND	£21
Upperud	£19
Liljendal, Thurbo and Svana	£17

This list of prices is interesting in that it shows quite clearly a new feature, which seems to have arisen with the decision in Britain to reduce the import duty on iron bars by some £5.10.0 per ton in 1826. The most surprising effect of this was a rise in the prices charged in Stockholm for the Walloon irons but not for the cheaper Swedish grades. The prices for the Russian irons were also unaffected. The Swedish records show this quite clearly.[33] Gimo and Ranas irons, having varied from 18 to 22 riksdaler per skeppund between 1817 and 1826, rose to 35 to 39 riksdaler in the period 1827 to 1829; similar figures were quoted for Osterby iron. Concerning the competitive nature of the Russian iron under these circumstances, two interesting communications have survived in the form of letters from Jonathon Marshall who, after the Huntsmans, was the most reputable Sheffield steelmaker of the early nineteenth century. On 16 July 1828 he wrote to Baron Tamm pointing out that the CCND Russian iron had been reduced in price since the duty had been lowered and, at the same time, had improved in quality, whilst the Oregrund iron was still the same price as previously. Then on 22 September the same year he wrote:

> 'I am sorry to say we have opposition in the trade in Sheffield at present and, as the CCND Old Sable iron from the late Count Demidoff's mines has been much improved these several years lately and is sold at 14 per ton less than the OO iron, much use is made of it for steel. I have always been fully convinced of the superior quality of the OO iron, having made use of that mark for more than 100 tons a year for the last twenty years, and should it be agreeable to you to order it made full half inch only and not thicker, I should be very glad and it would suit the whole trade much better.'

Year-end stocktaking lists from the Doncaster concern for the years 1834 to 1840 confirm the growing importance attached to the 'common' grades of Swedish iron and the increase in the amount of Russian iron used, three Russian marks appearing along with those from Upperud, Backefors, Svana and others which cannot be identified from the somewhat undecipherable representations of the various marks.[34] The Walloon irons still seem to predominate, however. There is a Huntsman stock list for 1843; of the 176 tons listed some 27 tons is Russian CCND and the remainder is entirely Walloon iron—45 tons Leufsta, 53 tons Gimo, 16 tons Osterby, 10 tons Forsmark, 5 tons Stromsberg, 11 tons Soderfors and 9 tons Harg.[35]

In the light of other isolated evidence from Sheffield steelmaking concerns, the preferences for the various types of iron have some significance. There are records of iron purchase by Daniel Doncaster and Company for the years 1851 to 1859.[36] There is also an iron purchase book from Tyzack and Sons for the years 1840 to 1868.[37] A further chance survival is the stocktaking register, with full valuations, from Thomas Firth and Sons for 1854.[38] The overall pattern from these last four sources is summarized in Table 1.

The four businesses, all operating within the Sheffield area, had different origins and different outlooks. Benjamin Huntsman and Company carried on the traditions of the founder of the crucible trade in Sheffield and their accent was on quality; they followed their old traditions and considered that only the best raw material would protect their well-established reputation. Tyzack and Sons were specialists in edge tools and cutlery and also followed the old traditions, although for some of the coarser applications—agricultural tools, for instance—a lower grade of material would suffice; none the less they seem to have chosen their sources of bar iron on the basis of experience. Those from Storfors, Backefors, Hammarby, Killafors, Dadran, Lesjofors and Ransater—most of them the so-called 'Lancashire' irons—were repeatedly called for, together with the three standard Russian

Table 1 Sources of iron purchased by four Sheffield firms during the mid-nineteenth century

Firm	Year	Total iron	% Walloon	% Common	% Russian
Huntsman	1843	176 tons	84.7	Nil	15.3
Tyzack & Sons	1840–47	336 tons	46.7	31.5	21.8
	1848–54	612 tons	47.4	32.4	20.2
	1855–61	998 tons	41.3	24.2	34.5
	1862–68	1,291 tons	40.0	34.2	25.8
	TOTAL	3,237 tons	42.5	30.5	27.0
D. Doncaster	1851–53	1,459 tons	29.0	61.6	9.4
	1854–56	1,760 tons	34.6	60.5	4.9
	1857–59	2,493 tons	30.0	46.6	23.4
	TOTAL	5,712 tons	31.1	54.8	14.1
T. Firth	1854	2,432 tons	13.7	86.3	Nil

grades. It is also significant that the second-grade Walloon irons, particularly Harg, Vattholma and Gysinge, were purchased in far greater quantity that those from Leufsta, Gimo and Osterby.

The Daniel Doncaster organization was a complex one. Originally filemakers, they had now become suppliers of steel to the Sheffield trades (and eventually they were to take up the merchanting of Swedish irons themselves). Their trade was, therefore, more diverse than that of either Huntsmans or Tyzacks and they were really selling blister bar and also crucible steel to their customers' requirements. These would include top quality material for razors, surgical tools, clock springs and the like as well as relatively cheap materials for general engineering purposes. A marked contrast to the first two firms is provided by Thomas Firth and Sons. Some idea of their diversity of operation can be gleaned from the stocktaking value given to crucible steel ingots (obviously written down, since the bar produced by simple forging of the top grade material sold at £55 to £60 per ton):

£35 per ton: Steel for taps, dies, graving tools, razors, axes, turning tools, pistol cylinders, scythe blades, pen nib sheet

£32 per ton: Steel for clock springs, music wire, brace bits, adzes, swords, saw files, quality blades & forks

£28 per ton: Steel for hammers, cane knives, augers, small files

£26 per ton: Steel for common blades & forks, butcher knives, cross cut saws, large files

£22 per ton: Steel for common files, hoes, shovels, machinery parts

It is of interest, too, that Firths were beginning to produce engineering forgings at this time and it is most likely that these would be produced from the cheaper materials. They had stocks of Leufsta, Osterby, Forsmark, Gysinge, Carlholm and Harg iron; their stocks of common iron were much larger and the major items were from Backefors, Storfors, Gravendal, Edsvalla, Salboda, Svartnas, Strombacka, Korssa and Hedaker—again, the 'Swedish Lancashire' marks predominating.

Foreign observers of the Sheffield scenes also commented on the diversity of iron used in the town in the middle years of the nineteenth century. Professor Le Play, who visited Sheffield in 1836 and again in 1842, having discussed the requirements in terms of 'body', 'strength' and 'toughness', came to the opinion that the price the steelmaker was willing to pay gave the best available classification of the various irons; his well-known list shows the ten Walloon irons at its head, followed by a variety of Swedish Lancashire irons. The five most expensive (£35 to £28 per ton) were Leufsta/Carlholm, Gimo/Ranas, Osterby, Forsmark and Stromsberg/Ullfors. There followed the remainder of the Walloon irons at prices between £25 and £21 per ton (together with Norwegian iron from Oester Rusoer and iron produced by the Franche Comté process at Sorfors in the north of Sweden). The best Russian irons and the major Swedish Lancashire grades follow, priced at £19 to £16 per ton.[39]

Heljestrand was in Sheffield in 1846 and his classification shows considerable divergence from Le Play's in that some of the Lancashire irons are shown as being preferred to some Walloon irons. His list reads as follows:[40]

First Rank:	Leufsta, Carlholm, Osterby and Gimo
Second Rank:	Ranas, Elfkarleo [Alfkarleo], Soderfors, Gysinge, Wattholma, Forsmark
Third Rank:	Skebo, Backefors, Lesjofors and Sorfors
Common Irons:	Nyquarn, Stromsberg, Ullfors, Sanna, Engelsberg, Fagertsa, Farna, Westanfors, Dadran, Furudal, Fallbo, Horndal, Narn, Schisshyttan, Svartnas, Upperud, Brunsberg, Forsbacka, Hogfors

In 1857, Professor Tunner of Leoben classified the Dannemora irons in order of rank[41] (Table 2), to which has been added the prices paid by Daniel Doncaster and Company in Sheffield in 1863[42] for comparison.

Similar rankings of the steelmaking irons occur elsewhere. E.F. Sanderson, writing to Dr John Percy in 1863,[43] quotes them in the following order: Leufsta, Gimo, Osterby, Forsmark, Carlholm, Soderfors, Gysinge, Russian Old Sable.

The Doncaster price lists for blister steel, a number of which have survived from the period 1857 to 1895, all show a similar ranking.[44] The earlier ones also indicate the suitability of the various grades for various applications. The most onerous application, as would be expected, was 'shearing', i.e. reforging into shear steel. The 1860 list notes all the Walloon

Table 2 Rank order of Dannemora irons according to Tunner

Forge	Number of Walloon hearths	Mean annual production (tons)	Sheffield price in 1863 (per ton)
Leufsta	3	1,008	£32
Osterby	2	670	£30
Forsmark	2	448	£29
Gimo	1	336	£31
Ranas	1	196	
Harg	2	392	£26
Vattholma	1	308	£25
Ullfors	1	168	
Stromsberg	1	308	£29
Alfkarleo	1	224	
Soderfors	2	616	£23
Carlholm	1	140	£24
Gysinge	3	440	£24

irons as suitable for shearing, together with the cheaper irons from Kihlafors, Backefors, Russian CCND and Storfors. Other common irons are listed as suitable for melting or for rolling to springs, but not for shearing; these include the other Russian irons, together with Swedish irons from Lennarfors, Hofors, Svaneholm, Elfsbacka, Finspong and Hasselfors. Later lists indicate the irons from Wikmanshyttan, Brattfors, Mo and Molnebo are suitable for shearing and that the Svaneholm iron is also upgraded into this group.

This exploration of the cheaper irons obviously was carried on extensively by the Sheffield steelmakers. A copy of the 1845 'Stampel-bok' which was used by the firm of Marsh Brothers has survived, with interesting annotations.[45] The iron from Svaneholm is here confirmed as suitable for shearing, together with those from Lahnfors and Hellefors, whilst several others are noted as 'good sound iron' or 'good melting iron'. There are also some derogatory comments: iron from Thurbo was 'poor', that from Svartnas was found to be 'unsound', Axmar was 'not very good', Gammelstilla was 'poor with no body' and Matfors was 'fair when sound'. It is, of course, the Dannemora irons which consistently receive the comments 'best mark' or 'best second mark'.

The Second Half of the Nineteenth Century

There were far-reaching changes in Sheffield steelmaking during the second half of the nineteenth century. The coming of the railway in 1838 had opened up the development of a large area of land along the lower Don valley which had previously been outside the town boundary, and it was here that the large works, those of Cammell, Firth, Brown and Jessop, together with a number of smaller works, were to be built between 1845 and 1855. At the time of their establishment all were based on crucible steel production; they all had large capacity cementation furnaces to provide the required blister steel. These were followed in 1863 by the erection, at the eastern end of this valley, of the River Don Works by Naylor, Vickers and Company, the largest crucible steel works in the world at the time.

In addition, however, the rate of technological change was continually increasing. With the advance in understanding of the steelmaking process and particularly of the role of carbon, it was proposed as early as 1800[46] that steel could be made by putting bar iron and charcoal into the crucible. Trials with this procedure, however, proved abortive. The melting point of the bar iron was higher than that of the blister steel—by some 50–80°C, dependent on the carbon content of the latter—and the charcoal was only absorbed when the metal was molten. The attainment of this higher temperature proved too difficult with the existing furnaces. As early as 1816 Broling suggested that the combination of suitable proportions of bar iron and cast iron in the crucible would provide steel of any required carbon content;[47] indeed he implied that he was using this practice and would eventually provide a report on it, but no such report seems to exist, unfortunately. This practice was first suggested in Sheffield in 1839,[48] but again, there was a practical difficulty. In this case, however, it was not related to furnace operation: indeed the melting down of the cast iron would commence at around 1150–1200°C and, as the temperature rose, the bar iron would be progressively absorbed, thus cutting down the need for excessive superheat in the furnace. The only problem was that cast irons with the requisite low contents of both sulphur and phosphorus were not available to the Sheffield steelmaker.

Then, in 1854, the Swedish Government's prohibition on the export of cast iron from Sweden was removed. This was a most significant action and heralded considerable changes in the Sheffield operations. Not only was the blending of Swedish bar iron and Swedish cast iron now a practical possibility for the crucible steelmaker; it eliminated the costs and the delay involved in putting the bar iron through the cementation furnace. The Swedish cast iron could be utilized in other ways. In the first place, a number of the larger steelworks had installed puddling furnaces, essentially for the production of wrought iron for plates and forgings. The purchase of Swedish cast iron and its conversion to bar iron in the Sheffield puddling furnaces produced a much cheaper source of good quality iron than even the lowest priced Swedish bar iron imports—in 1861 John Brown was selling a 'melting base', which was such a product, at £13 per ton when the cheapest Swedish iron was £15 per ton and Leufsta iron was £36 per ton.[49] Moreover, the technique for the production of what came to be known as 'puddled steel', by controlling the carbon removal from the molten iron in the puddling furnace, became a common practice in these Sheffield furnaces in the late 1850s.[50] This was not only useful as a higher-strength form of wrought iron for engineering purposes but, when produced from the higher-purity Swedish cast iron, could also be used as the charge for the crucible process in place of blister steel.

The development of the gas-fired crucible furnace during the early 1860s, based on the work of Siemens, enabled operation at higher temperatures than those previously available from the simple coke-fired furnace.[51] It is interesting to note that this caused an alteration in the plans for the massive River Don Works of Naylor, Vickers and Company. It was originally designed to have a melting shop with 336 coke-fired double melting holes, so that 672 crucibles of steel could, if necessary, be available at any one time. When two-thirds built, the scheme was changed and the remaining capacity was installed as gas-fired furnaces. In these, more heat-resisting crucibles, manufactured from clay containing a substantial graphite addition, were used and the charges to the crucible were bar iron and the appropriate amount of charcoal, thus bringing the idea first tried in 1800 into operation. The remaining coke-fired furnaces were, it seems, used with mixed charges of Swedish bar iron and Swedish cast iron, thus fulfilling the anticipations of the earlier Vickers patent of 1839. Thus it was that the River Don Works, with its capability of 15,000 tons of crucible steel per annum, was operated without the inclusion of a single cementation furnace; this was a major change brought about by improved technology and raw material availability.

It was in 1856, of course, that Henry Bessemer announced to the world his revolutionary process of 'making iron and steel without fuel'.[52] The story of his early problems and eventual success is well documented elsewhere. What should be said, however, is that without the help of Goran Goransson at Edsken and the export of Swedish cast iron to Sheffield his success would have been much slower in achievement; whilst the importance of the Bessemer steelmaking invention cannot be denied, it is a pity that the man himself omitted to acknowledge his debt to the Swedish workers when he prepared his own story of the development.[53] Nevertheless, the successful development of the Bessemer process had some interesting effects on Sheffield steelmaking and its relations with Sweden, as will be considered later. The initial effect was that both John Brown and Charles Cammell added Bessemer converters to their works in the early 1860s, using cast iron derived from the Cumberland haematite ores, and thereupon entered the new business of making steel rails, destined to be used all over the world wherever new railways were being built.

Firths and Jessops, however, remained faithful to their established crucible steel operations, at least for the next two decades. There must have been some minor overall adjustment, in that the cheaper Bessemer steel would also take the place of the lower grades of crucible steel previously used for common applications in general engineering, but in the main this did not interfere with the old trade. Indeed the market for crucible steel tended to grow, since it was increasingly used for the production of forgings. Larger and larger ingots and, for that matter, larger and larger steel castings, were being made by combining the contents of several crucibles. Naylor, Vickers and Company in 1860 cast a steel bell for San Francisco fire station which weighed over 5 tons and required the steel from 105 crucibles;[54] the same firm cast an ingot of 25 tons in 1869 using the full capacity of the melting shop—672 crucibles full of steel.[55] Thomas Firth and Sons were using similar practices—an ingot for a gun forging in 1874 took 574 crucibles.[56] The technique was essentially simple: a tundish set in the floor of the melting shop would have its nozzle centred over the casting mould or ingot mould set in the cellar below. When the steel in all the crucibles was ready for casting, a simple plug of crumpled paper would be rammed into the nozzle and the contents of a number of crucibles quickly poured in to provide a reservoir of liquid before the paper burned through, letting the metal into the mould below; meanwhile, liquid steel was poured in continuously from other crucibles, so that a constant stream of metal flowed from the nozzle until the mould was full. The real difficulties in such an operation lay in marshalling the number of operators necessary to ensure that the crucibles full of steel were at the right place at the right time.

At the same time, however, these large firms were still carrying on their old businesses based on the simple small ingots cast from single crucibles and the forging down of these to provide material for the cutlers, the tool makers, the file makers and the other local craftsmen, as well as supplying an ever growing export business, since most of the Sheffield steelmakers had agents on the Continent, in America and in the British colonies. New firms also entered the trade. Henry Seebohm, a partner in one of the most successful of these—Seebohm and Dieckstahl, later to become Arthur Balfour and Sons—in a lecture given in 1868 pointed out that there were then 150 businesses in Sheffield engaged in the making of steel and stated that a shop with twelve melting holes would provide sufficient return to its owner to afford the luxury of a carriage and pair, then the top status symbol! More importantly, however, he had much of interest to say concerning the manufacture of 'Cast Steel' (an alternative name for crucible steel at the time) and the value of Swedish iron.[57]

Seebohm classified 'Cast Steel' into three categories according to the origin of the iron. English Cast Steel he wrote off as inferior, being made from native English iron together with spring ends, Bessemer scrap and the like—all cheap material and, by the time he wrote in 1868, largely superseded by common Bessemer steel. The second category, which he referred to as Swedes Cast Steel, was made from ordinary Swedish irons. These he put into three groups. The lower rank he implied had never succeeded in attaining a good reputation as melting irons, providing material which was relatively cheap but which performed a useful function for common quality saws, second quality files, miners' drills for soft stone and other tools where the finishing cost was small compared with the cost of the steel; the steel was lacking in 'body' and was unfit for turning tools or other tools where a sharp cutting edge had to be maintained. The grades of iron were not specified but were stated to have had an average price in Hull of £14 per ton from 1846 to 1868. His next group was based on irons stated to have had an average price of £16 to £18 per ton over the same period, irons much superior to the previous and including those from Bjurback, Bjorneborg, Svabensverk, Gammelstilla 'and, perhaps, a hundred others'; filemakers with a good reputation were said to buy this class of steel. The top rank of these irons were stated not to be so numerous, including Soderfors Lancashire iron, that from Killafors 'and a few other less well known marks' which had an average price of £21 per ton; he did not specify the uses of such materials. His third class was Dannemora Cast Steel. 'The experience of many years, at least a century, has proved that these irons make a better steel than any other

Swedish or other marks.' He classified these into three groups:
1. Common Dannemora Marks: C and Crown (Alfkarleo), Gridiron (Wattholma), Little S (Skebo) and Steinbuck (Harg)
2. Middle Dannemora Marks: Hoop F (Forsmark), JB (Gysinge) and W and Crown (Stromsberg/Ullfors)
3. Best Dannemora Marks: Double Bullet (Osterby), GL (Gimo/Ranas) and Hoop L (Leufsta).

> 'Each of these marks has its own admirers and some are especial favourites for particular purposes. Hoop L is the mark best known in this country. If a country blacksmith wants a bar of Blister Steel for welding he gives preference to this mark above all others. It is also one of Huntsman's favourite marks for melting and is largely used by Rodgers, the most celebrated cutlers in the world, for shearing for their best table cutlery. GL is another—we might almost say the other—mark of which Huntsman makes his celebrated Cast Steel. W and Crown has the reputation of making a very hard Cast Steel and is largely used by Spencers, Hall and other celebrated saw file makers. Steinbuck is a very favourite mark for making Shear Steel, especially for table blades.'

Seebohm attributed the superiority of these top-class irons for Cast Steel manufacture to their relative freedom from impurities compared to the common irons and their small percentage of phosphorus, sulphur, copper and silicious matter. He goes on to comment:

> 'That they make better Cast Steel than the other marks of Swedish or Russian irons may be inferred from the fact that the Sheffield Steel Manufacturers have been willing to pay such a high price for them for so many years.'

Table 3 shows Seebohm's record of prices per ton in Hull as examples in each group.

Table 3 Price per ton of the top-grade irons

Mark	1847–51	1852–3	1854	1855	1856–63	1864–6
Steinbuck	£23	£22	£24	£26	£26	£24
W and Crown	£24	£26	£28	£32	£30	£29
Hoop L	£32	£29	£31	£36	£34	£32

His further comments are enlightening:

> 'The great competition in the Steel trade has induced many houses to try cheaper marks of which to make best Cast Steel. So called Best Cast Steel has been sold at prices at which it is obvious that much cheaper marks than the Dannemora marks must have been used; but none of the cheaper houses have met with any permanent success and

> experience continually proves that nothing is so dear as cheap steel
> ... If an engineer wants good Cast Steel he must be willing to pay a
> price for it sufficient to allow of the best marks being used ... One
> thing may be learnt from a due comprehension of the great variety of
> Cast Steel that may be made, and that is the importance, which cannot
> well be overestimated or too often insisted upon, of stating when
> ordering Cast Steel the purpose for which it is required.'

He further pointed out the variety of methods of making the steel, as distinct from the choice of iron involved, and set down four different routes to a melt of 1% carbon steel:

- A. By putting into the melting pot Blister Steel converted to a No. 4 temper, which is known to contain about 1% carbon.
- B. By putting into the melting pot unconverted Iron and charcoal in the proportion of 50 lbs [22.7 kg] iron and 8 oz [227 g] charcoal.
- C. By putting into the melting pot unconverted Iron and Spiegel Iron in the proportion of 40 lbs [18.1 kg] iron and 10 lbs [4.5 kg] Spiegel Iron containing 5% carbon.
- D. By putting into the melting pot Blister Steel converted to a No. 5 temper and mild Cast Steel scrap in the proportion of 34 lbs [15.4 kg] of the former, which contains 1.25% carbon, and 16 lbs [7.3 kg] of the latter, which contains 0.5% carbon.

He continued:

> 'Whatever may be the difference in the chemical composition of these
> various kinds of Cast Steel their properties are different ... the
> following is a correct order of merit, placing the most meritorious
> immediately after the special good quality which it exhibits:
>
> | | |
> |---|---|
> | Welding easiest | BCAD |
> | Combining the greatest amount of hardness and elasticity when hardened | ADBC |
> | Combining the greatest amount of hardness and toughness when unhardened | CBAD |
> | Smallest tendency to honeycomb in the melting (casting) | CDBA |
> | Smallest tendency to water crack in the hardening | DBAC' |

With more and more forgings and castings being produced, mainly from crucible steel, but with the old-established small crucible steel trade still thriving, it became clear that the Sheffield steel trade was really developing on two different lines, although both were still dependent on crucible steel and Swedish iron. There was an anomaly, however, in the rail trade based on the Bessemer converter. This was to change; the year 1873 saw the start of what the economic historians usually refer to as 'The Great Depression'. This was the start of a rapid decline in the rail trade; by 1875 the smaller firms involved in it had gone out of business, John Brown and Company were concentrating on engineering products and only Cammell continued their railmaking activities. In 1883 they too removed their complete rail-making plant to Workington and there was virtually no Bessemer activity left in the Sheffield area.

By this time, however, another factor had to be taken into consideration. The open-hearth furnace had by now been established as a reliable means of producing larger quantities of good-quality steel without the necessity for combining small-scale melts to provide a forging ingot or a casting weighing several tons. From about 1879 onwards this process took over in what might be termed the 'heavy' Sheffield steel trade. This change in melting practice was accompanied by the installation of large hydraulic forging presses and heavy mills; by 1890 plates, forgings and castings for ordnance, marine and general engineering purposes occupied much of the capacity in the Sheffield works. Surprisingly enough, the open-hearth steelmelting was all in acid-lined furnaces; the basic variants, although well established by 1890, were not considered to give a product of sufficient quality, whilst Bessemer steel would not even be considered, either acid or basic. So it was that Swedish cast iron and Swedish iron ore, low in sulphur and phosphorus, became valuable to Sheffield's forging trade.

The lighter trade, based on the old small-scale crucible melting operations, still continued and the new engineering giants of Firths, Browns, Vickers and Jessops all continued their well-established and rightly celebrated brands of crucible steel, made in the old traditional ways, alongside their new developments. Indeed, the requirements of the machine tool trade grew with the expansion in engineering and the greater quantities of metal removed in the massive machine shops; the growth of special tool steel production, which started with 'Mushet's Self Hard' in the late 1860s, became a feature of all the Sheffield firms at the end of the century. Meanwhile there were the smaller firms which continued single-mindedly with their crucible melting shops, many of them buying in blister steel from major suppliers such as Doncasters, who were still issuing their blister steel price lists in 1895—and presumably continued to do so, although no lists seem to have survived.

There was, however, a most interesting further development from about 1880 onwards. The lack of confidence in Bessemer steel for other than 'common' applications has been remarked on earlier. Swedish Bessemer steel, however, was looked upon differently, and two Sheffield firms, Daniel Doncaster and Sons and George Senior and Company, began to operate a trade in Swedish Bessemer steel. This was originally obtained in billet form and worked down in Sheffield into bars, which were then sold to the trade for general engineering purposes. Subsequently, the import of ingots started, in a range of carbon contents from 0.65% to 1.30%; these were forged down in Sheffield and then processed to sizes of bar called for by the growing range of customers. This steel had a reputation for regularity in heat treatment response and was, it seems, much sought after on account of its good quality coupled with its reasonable price. Moreover, the discards from the bar production often found their way into crucible melts: it was good material from this point of view, low in sulphur and phosphorus and with a known range of carbon analyses.

The Twentieth Century

The opening years of the twentieth century witnessed the development of high-speed tool steel as a development from the Mushet 'Self Hard'. Mushet's original work in 1868 involved the melting of Swedish white cast iron together with wolframite ore in crucibles, the resulting material containing up to 2% carbon, with 6-10% tungsten and 1-2% manganese; he later incorporated chromium ore also, giving up to 1% chromium. Crucible makers worldwide were producing similar types of material by 1900, and the composition was being modified, largely as a result of the work of Taylor and White in America.[57] The well-known 18-4-1 analysis (18% tungsten, 4% chromium, 1% vanadium, with a carbon content of around 0.7%) became the normal standard from about 1910 onwards.

A few examples of charges used for these tool steels have survived; they invariably incorporate Swedish iron as raw material. One from a small works with four crucible holes[59] used Swedish blister bar, box ends and white iron, no brands being specified, as base charge in making a 'Self Hard' in 1897. Two early melts of high-speed tool steel from William Jessop and Company[60] quote the use of blister steel from DU iron (Lancashire iron from Stjernsund) or 'Best Blister', which would undoubtedly be from a Walloon iron. There is an interesting comment just after the First World War by an American observer[61] that, despite tests carried out during the war which showed that perfectly satisfactory tool steel could be made using Armco iron or specially purified domestic iron, the exclusive use of Swedish iron was taken up again after the war as soon as supplies were readily available.

Carbon tool steels were also still being made in considerable quantity and those designed for heavier use often had 3-4% tungsten added; there is a reference to 'EXTRA' being half Double Bullet and half W and Crown, whilst 'DOUBLE EXTRA' was half Leufsta and half Gimo.[62]

Variations in the cost of Leufsta iron can be traced through the Doncaster records[63] for the first quarter of the century, the figures being for prices in shillings and pence per hundredweight in Sheffield:

1902-5	1908	1911	1913	1914	1916
19/6	18/6	18/3	17/6	18/6	36/0
1919	1920	1922	1923	1925	1926
49/0	54/0	30/0	30/0	29/0	28/0

There is some valuable information in the charge books for the crucible furnaces at Charles Cammell and Company for 1925-28.[64] Much of the material melted was carbon tool steel, together with considerable quantities of various grades of high-speed steel; there was also a significant proportion of alloy engineering steels together with a small quantity of what might be called 'specials', including 36% nickel steel and a number of the stainless and heat-resisting steels which were then being developed. The crucible process proved ideal for development work with small experimental quantities. The general pattern was the use of 15-30% return scrap together with the necessary ferroalloys, the remainder being Swedish iron. Very little of the top-grade Walloon iron was used in alloy steel, although Leufsta and

Gimo was incorporated in many of the special grades of carbon tool steel. The Swedish irons quoted were AOK—the largest single brand used—together with Crown and Anchor, DU, DGL and MR Box Ends. Some 1925 analyses and prices are listed in Table 4.

Table 4 1925 analyses of four iron grades

Grade	C	Si	Mn	S	P	Price per ton
Hoop L	0.10	0.05	0.28	0.009	0.011	£37.10.0
DGL	0.08	0.03	0.16	0.007	0.011	£29.0.0
AOK	0.04	0.01	0.15	0.015	0.017	£17.10.0
MR Box Ends	0.05	0.02	0.04	0.008	0.030	£12.15.0

Three years later the prices per ton for various irons were quoted as follows:

Hoop L bar	£35.0.0
Hoop L ends	£24.0.0
GL ends	£24.0.0
AOK bar	£16.0.0
AOK bar, converted	£20.15.0
Crown and Anchor bar	£14.5.0
MR Box Ends	£12.12.6
Swedish White Iron	£9.0.0

One further source of evidence for the use of Walloon iron takes us, in effect, full circle, since it not only comes from Benjamin Huntsman's own firm[65] but stresses the value still put by them on the old tradition of using only the best. In the eleven years from 1924 to 1934 their blister bar purchases from other Sheffield firms—they had abandoned their own cementation furnaces when they moved to new premises in 1898—totalled 1,397 tons. Of this 654 tons was Hoop L and 529 tons GL, still bearing out what Seebohm had said about them in 1868; the remainder were cheaper Walloon grades, mainly W and Crown and JB. In addition they purchased some 27 tons of common Swedish grades of iron, including LTS, RW^n and DU. They also bought return scrap amounting to 223 tons and used 26 tons of alloys. Their total intake, therefore, was 1,673 tons over the period 1924 to 1934, some 84% of which was Walloon iron—virtually the same proportion as in 1843!

One intriguing feature is that purchases of Hoop F figure in the records for 1931 to 1933; it is, of course, known that Forsmark ceased production of bar iron in 1897. The reason appears to be that Strombacka bought the Forsmark share in the Dannemora mines in 1881[66] and marketed iron with the 'Hoop F' stamp for many years thereafter. The reported use by Whitworth, the Manchester engineering firm, of 700 tons of Hoop F, together with 900 tons of JB and 400 tons Double Bullet on 'special work', around 1900, could, therefore, involve material not from Forsmark but from Strombacka.

The final piece of information concerns the use of iron in the cementation furnaces at Daniel Doncaster from 1918 to 1951.[67] Over the period 1918 to 1940 the charges were entirely of Swedish iron. The marks listed are Hoop L, AOK, SL, DGL, Little S, W and Crown and LTS. From 1940 to 1950 some 85% of the 2,571 tons converted was English iron, specially made for the purpose by Low Moor and designated 'LMAS'. Included, however, were some 87 tons of Leufsta iron and 20 tons of Gimo iron, plus 163 tons of other Swedish irons. The historic last heat, charged on 27 October 1951 in the No. 2 Doncaster furnace—believed to be the last production of blister steel in the world—is intriguing in that it contained 14.1 tons of Hoop L iron, almost a quarter of a century after the forge at Leufsta had closed! There was also one ton of PAT, 8.2 tons of Swedish Bessemer and 8.2 tons of LMAS. As against a conversion cost of around £1 per ton in 1870–80, this last heat, perhaps not surprisingly, cost almost £12 per ton.

HULL IRON MERCHANTS AND SHEFFIELD STEEL

The trade in Swedish iron between Hull and Sheffield can be traced back to the early years of the eighteenth century through the records of the Fell 'Steele Trade'.[68] These show quite clearly that the Hull merchant house of Sykes was supplying Sheffield with Walloon iron from Leufsta as early as 1701, whilst iron from Leufsta, Osterby, Akerby, Alfkarleo, Vessland and Harg was supplied by 'Mr Sykes' during the period 1712 to 1724. Some material in these years was transferred to Hull from London; the merchants named were Mr Parkin and Mr Boughton, and in 1720 there was a transhipment from Holland via Mr Stallard. Other regular suppliers, all dealing exclusively in Walloon irons and mainly in those from Alfkarleo and Akerby, were Mr Thornton, Mr Victorin, Mr Mowld, Mr Fenwick and Mr Wilberforce. Apart from the house of Sykes, which continued as a major supplier of the highest quality irons until the middle of the nineteenth century, none of these names can be traced in the surviving records in Hull. It must, however, be more than mere coincidence that the list of mayors of Hull shows William Mowld holding the office in 1714 and again in 1734, with William Wilberforce in 1722, William Fenwick in 1727, Joseph Sykes in 1761, 1777 and 1792, and Thomas Mowld in 1764. It is a reasonable assumption, therefore, that most of the supplies for the Fell 'Steele Trade' came through Hull; what is clear beyond any doubt was that the iron used in Sheffield from 1701 to 1765 was all 'Dannemora' or 'Oregrund' iron from the Walloon forges in Uppland.

One hundred years later, the surviving Sheffield records provide direct evidence of the Hull–Sheffield trade in Swedish iron.[69] This can be supplemented by the Hull Customs Bills of Entry.[70] A survey of these has been made for a number of years between 1832 and 1852 and the results have been summarized in Table 5, arrived at by integrating all the various items according to port of origin of the irons imported into Hull. It is interesting to note that the amounts of Russian and Norwegian iron, within fairly broad limits, tended to remain constant, whilst there were large fluctuations in Swedish iron imports. The other feature which becomes clear was the rise in importance of Gothenburg as an iron-exporting centre, supplying more

Table 5 Imports (tons) of iron bars into Hull

	1832	1837	1840	1842	1845	1847	1849	1852
Swedish iron								
From Gefle	1,305	776	2,490	2,096	2,905	3,097	1,148	2,854
From Stockholm	5,008	5,007	3,476	4,455	11,417	11,328	7,290	7,947
From Gothenburg	711	674	1,256	1,762	6,752	6,211	7,196	8,696
From other ports	121	57	50	89	323	795	628	1,225
TOTAL	7,145	6,514	7,272	8,402	21,397	21,431	16,262	20,722
Norwegian iron	282	462	289	269	470	299	349	437
Russian iron	1,216	1,652	1,615	1,619	1,812	1,709	2,825	1,580
Total bar iron imports	8,643	8,628	9,176	10,290	23,679	23,429	19,436	22,739

than Stockholm by 1852. Gefle at this time appears to have had a limited capacity of around 3,000 tons per annum, and in 1852 the steady rise of exports from the Baltic ports north of Gefle actually totalled more than had been exported from Gefle in 1849.

The main value of the Customs records, however, lies in their identification of the merchants who took over the iron in Hull. Table 6 indicates the major importations of iron bars as revealed by these records. Most of the merchants, including Sykes, Wilson, Spence, Wilkinson, Clay, Gee, Graham and Hewitt, appear in the purchase ledgers of the Sheffield firms. On the other hand, there is nothing known of Laverack, Pauling, Lofthouse and Gardam, Earle & Co., nor is there any entry for any of them as Iron Merchants in the Hull Trade Directories for the period; they may, therefore, have been firms based elsewhere in the country, using Hull as convenient for the North Sea trade.

The imports of Russian iron in the period 1832 to 1842 were largely to the account of Edward Spence; in 1842 and 1845 the Graham concern handled fair quantities, but the major importer from 1847 onwards was the Wilkinson firm. As far as Norwegian iron was concerned, the main agent here throughout the period studied was Clay and Company; it is significant that Joseph Sykes, best known as the main supplier of the top-grade Walloon irons, actually imported small quantities of Norwegian irons in both 1840 and 1842. The firm of Joseph Sykes and Company had, indeed, obtained a virtual monopoly of the product from the Leufsta, Osterby, Gimo and Ranas forges as early as 1760 and held this for almost a hundred years. In the 1830s the bars from these sources were actually stamped with the name 'SYKES' as well as with the appropriate forge stamp[71] and it is clear from the few surviving Sheffield records that no Hoop L, GL or Double Bullet iron was supplied by any other house until after the middle of 1854. It will be noted from Table 5 that the quantity imported by Sykes and Company remained reasonably constant from 1842 to 1852, despite the dramatic increase in the overall total of Swedish iron which came into Hull, underlining the specific nature of the Sykes business and the relatively small amount of the best quality Walloon iron available.

Table 6 Imports (tons) of Swedish bar iron into Hull by main importers

	1832	1837	1840	1842	1845	1847	1849	1852
Joseph Sykes & Co.*	4,231	2,711	1,412	2,356	2,629	2,262	3,217	2,654
Beckington, Wilson & Co.*, later Thomas Wilson, Sons & Co.*	1,100	1,134	1,753	1,688	7,701	4,164	2,897	2,637
Edward Spence *	123	339	871	1,044	983	942	411	994
Wilkinson, Whitaker & Co.*	590	1,540	1,805	1,542	4,542	5,686	6,387	5,311
Gee, Loft & Co.*	357	153	129	501	528			
W. Laverack	344		416	714	1,268	2,664	1,104	
Clay & Co.*, later Clay & Squire *	119	337	453	264	1,311	2,701	8	19
Barnby & Co.					967			
H.D. Pauling						1,548	825	879
D. Lofthouse & Co.					331	451	527	1,187
T.F. Hewitt *								709
J. Walker & Co.								1,004
Gardam, Earle & Co.								3,528
Good, Flodman & Co.*								343
Percentage of total Swedish bar iron imports into Hull handled by above importers	96.1	95.4	94.0	97.0	94.7	95.2	96.3	93.0

*Signifies known supplier to the Sheffield steel trade.

The background to some other iron importing businesses in Hull can be traced.[72] The important firm of Wilkinson, Whitaker and Company seems to have been a successor to William Williamson, who was known to have been trading in the eighteenth century. They are known from the Sheffield records from the 1830s as suppliers of the second rank of Dannemora irons—Harg, Wattholma and Stromsberg—together with the major Russian irons, as well as increasing amounts of Swedish Lancashire iron, particularly that from Backefors. Between 1853 and 1856, both partners died and the firm continued under the name of Fewster Wilkinson and Company, eventually ceasing business in the 1870s. The firm of Beckington, Wilson and Company, founded in 1825, with Thomas Wilson as the resident Hull manager, rapidly grew in importance; by 1838 the name had changed to Wilson, Hudson and Company and from 1841 until the retirement of Thomas Wilson in the mid 1860s the firm was styled Thomas Wilson, Sons and Company and carried on a very profitable business in Swedish Lancashire and Russian irons. Thomas Wilson also became a ship owner; many years after his death his successors entered into a partnership which brought into being the Ellerman-Wilson Line.

The end of the Sykes dominance in the Walloon iron trade came with the withdrawal of the last of the Sykes family from the business; the operations were taken over in part by T.F. Hewitt, who had been working as manager for the Sykes concern, but the contracts with the Walloon forges had run their course. There is a note[73] dated 27 May 1854 to the effect that no further supplies would come from Joseph Sykes and Company; in future Naylor, Vickers and Company would supply Double Bullet whilst Wilkinson, Whitaker and Company would supply Hoop L and GL. This, indeed, brought in a profound change in the Hull iron trade, which was commented on by other observers. The French, for instance, noted that the Sykes monopoly had only in part been taken over by the two firms and that the English steelworks no longer had the exclusive use of the premier iron grades, to the benefit of French and American steelmakers.[74]

It had been assumed that useful corroborative evidence might be gained from the Customs records for this period; unfortunately only fragmentary copies have survived from 1853 to 1857.[75] What is available does show that the last delivery of material for the account of Joseph Sykes and Son was in February 1855; in 1857, Fewster Wilkinson and Co. made major importations from Stockholm late in the year, as had been the pattern of the Sykes entries in previous years, and these may well have been the trade in Walloon iron. The incompleteness of the record for these years makes any real comparison with earlier years impossible. It has to be noted, however, that there was an isolated shipment of 101 tons of iron from Gefle—presumably of Osterby iron—to the account of Naylor, Vickers and Company in 1855; this is the only reference found in these records to this particular merchant house. In the same year there is an entry showing a shipment of 319 tons of iron from Oregrund, the first time that this port figures in these records. Another feature worth comment is that the imports of Russian iron temporarily cease in 1854 and 1855, presumably as a consequence of the Crimean War.

To compensate in some measure for the lack of data in the Customs records for this critical period, some valuable information can be gleaned from the authorization dates of the additional stamp marks on the various irons.[76] On 13 July 1854 the stamp of 'N.V. and Co' (representing Naylor, Vickers and Company) was authorized on Osterby iron; on 6 November 1854 'WILKINSON' began to be stamped on iron from Gimo. These substantiate the changes noted above. There is no similar evidence concerning the Leufsta iron; nevertheless, whilst other Sheffield ledgers indicate that Wilkinson, Whitaker and Company (later Fewster Wilkinson and Company) supplied Hoop L iron on a regular basis from 1854 to 1864,[77] the Doncaster records indicate material with an additional 'WILKINSON' stamp being offered during the period 1863 to 1866 by Sheffield merchants.[78] This could well have been a sale of surplus material purchased previously, however, since in 1864 there was an interesting sequence of events. In the first instance, the additional stamp of 'HINDE AND GLADSTONE' was authorized from 18 May, whilst a further amendment adding the name 'LEUFSTA' to the Hoop L stamp was recorded on 28 September 1864. This was supplemented by a statement from the firm of Hinde and Gladstone, of London, dated June 1864,[79] who indicated they were now the sole contractors for the Swedish Hoop L Steel Iron and 'being determined to adopt every practicable mode of protecting the trader in and consumer of the above named iron, we give this public notice that no Leufsta iron . . . which will be hereafter manufactured will be sold or delivered by us without the additional mark LEUFSTA being impressed thereon, such additional mark having been adopted by us for the further protection of the public and ourselves.'

This, incidentally, was not the only occasion that steps were taken to prevent the fraudulent use of the Hoop L stamp mark; in the late 1870s Baron de Geer took legal action in the British courts to prevent the use of the mark on material produced at the Brades Works, near Birmingham, the eventual decision being in favour of de Geer 'in that such a transaction would be calculated to deceive the purchaser'. He subsequently thought fit to issue a public statement complaining of the counterfeiting of his trademark on iron not produced at Leufsta. Hinde and Gladstone, however, do not appear to have remained sole concessionaires for Leufsta iron for much more than a year, since supplies into Sheffield in 1866 were from N. and M. Hoglund, either direct from Stockholm or via their agent, the London merchant W.C. Taylor, who was also dealing in both Gimo and Osterby irons.[80]

From 1858 onwards the Hull Customs records have, in the main, survived and these have been studied in detail for selected years between 1858 and 1873;[81] the information is summarized in Tables 7 and 8. It will be noted that the pattern is completely different from the previous position, in which most of the merchants were also identifiable in the Sheffield steel-

Table 7 Imports of iron bars and pig iron into Hull

	1858	1862	1866	1869	1873
A. IRON BARS					
Swedish iron					
From Gefle	3,683	7,520	6,920	7,397	10,108
From Stockholm	5,367	6,195	6,218	3,183	5,470
From Gothenberg	8,143	7,256	11,975	9,553	12,438
From other ports	2,422	1,079	2,283	5,584	5,842
TOTAL	19,615	22,050	27,396	26,017	33,858
Norwegian iron	414	67	136	331	Nil
Russian iron	991	5,552	2,613	84	1,332
TOTAL BAR IRON IMPORTS	21,020	27,669	30,145	26,432	35,190
B. PIG IRON					
Swedish pig iron					
From Gefle	50	857	332	459	831
From Stockholm	Nil	969	Nil	107	2,782
From Gothenberg	2,111	3,453	779	3,523	4,744
From other ports	Nil	614	91	327	1,890
TOTAL	2,161	5,893	1,202	4,416	10,247
Norwegian pig iron	Nil	36	233	788	Nil
Russian pig iron	Nil	331	Nil	Nil	20
TOTAL PIG IRON IMPORTS	2,161	6,260	1,435	5,204	10,267

Table 8 Imports (tons) of Swedish bar iron into Hull by main importers

	1858	1862	1866	1869	1873
Edward Spence*	1,997	484	593	2,126	
Fewster Wilkinson & Co.*	2,769	3,563			
Edward Squire, Jnr.*	590	639	1,101	10	
H.D. Pauling	318	635	126	704	194
Hewitt and Pease*, later T.F. Hewitt & Co.*	413	114	135	67	140
Gardam, Earle & Co.	2,121				
Earle, Woodall & Earle		7,669			
Earle, Haller & Earle			14,359	10,295	18,200
Good, Flodman & Co.*	79	1,437	1,059	367	440
G. Malcolm & Sons	3,547	2,728	269	605	559
H. Denniss & Son	2,147	1,625	2,143	1,477	1,370
Sahlgreen & Carrall	175	100	1,090		
Helmsing & Son			1,014	13	
E. Thompson & Son		125	288	358	385
T.H. North					796
M.S. & L. Railway					533
W.N. Smith					471
J. Bilton & Co.					399
Percentage of total Swedish bar iron imports into Hull handled by above importers	89.7	93.3	97.1	98.9	98.0

*Signifies known supplier to the Sheffield steel trade.

works ledgers. Now, apart from Wilson, Spence and Wilkinson, the names of the major Hull importers are not known in Sheffield. They could, of course, have been intermediaries, selling to Sheffield via third parties. Over the period studied, the importations are increasingly in the hands of the Earle family, first as Gardam, Earle and Co. (1858), then as Earle, Woodall and Earle (1862) and eventually as Earle, Haller and Earle (1866 onwards); the latter firm does eventually appear as 'Iron Merchant' in the 1872 Directory. It has to be admitted that the data collected from these records throw little light on the specific question of the supply of Walloon iron to Sheffield; at least some of this came via London, it being known that Hinde and Gladstone and W.C. Taylor and, very likely at this date, Naylor Vickers, operated from the capital as purveyors of Walloon iron. In 1858 and 1862 Fewster Wilkinson and Co. continued to import solely from Stockholm in quantities which are reminiscent of the Sykes trade; they too then vanish from the records. There are, however, some interesting comments to be made.

Firstly there is evidence of the growing trade in Scandinavian pig iron, first released in 1855, and details can also be found in Table 7. By 1873 it represented some 20% of the total iron imports through Hull. This particular year is interesting in that it is quite clear that it witnessed the peak output of steel by 'the old Sheffield methods'—cementation and crucible steelmaking. The available information leads to an estimate of some 120,000 to 130,000 tons of steel in the year by these means. Hull could provide 35,000 tons of bar iron and 10,000 tons of pig iron; it seems that London could have added a further 30,000 to 35,000 tons of bar iron.[82] With pig iron in the same proportion, this would yield between 80,000 and 90,000 tons of Swedish raw material; with added scrap and, perhaps, a modicum of domestic iron from haematite ores, the postulated output from Sheffield is not impossible, leaving some imported Swedish pig iron for use in the new Bessemer method of steelmaking. A point which should be stressed in such a context is that less than 10% of the Sheffield output would be in the superfine quality provided by the use of Walloon iron.

The other very noticeable feature, however, is the increasing amount of iron shipped from the ports on the eastern seaboard of Sweden, excluding Stockholm and Gefle. Indeed, by 1869 and 1873, something of the order of 20% of the total Swedish iron imported into Hull came from such ports—Kalix, Lulea, Skelleftea and Pitea in the far North, Hernosand, Sundsvall, Hudiksvall and Soderhamn in the central region, through Nykoping, Norrkoping, Vestervik and Kalmar to Karlscrona in the South. The interesting outlets with regard to Walloon iron, however, could be Skutskar, Oregrund, Hargsbruk itself, Osthammar, Grislehamn and Nortalje; in 1869 these ports jointly exported some 1,168 tons of bar iron to Hull; the figure in 1873 had risen to 2,082 tons.

From 1867 onwards, however, the only detailed information on the Walloon iron trade which is still available comes from the Daniel Doncaster and Company 'Contract Books', in which were recorded offers of various parcels of irons and also their own purchases; occasionally details of what was going on elsewhere in Sheffield are noted.[83] In 1867, it seems, a further change in policy occurred, in that the old-established Sheffield steelmakers took direct steps to ensure that their supplies of the top-rank Swedish irons should remain available to them. Naylor, Vickers and Company, of course, had become merchants in addition to their steelmaking interests some thirteen years earlier, but their merchanting was now in the hands of a separate organization, Naylor, Benzon and Company, whose interests lay more in America than in Sheffield. The first in the field was the firm of William Jessop and Sons, who had their stamp authorized on Leufsta iron on 23 May 1867, adding both Osterby and Forsmark irons in 1869, whilst Sanderson Brothers made similar arrangements with regard to Gimo iron commencing 7 June 1867.[76] Although no additional stamp was authorized, it is clear that Thomas Firth and Sons acted similarly, since by 1869 they were selling the irons from Soderfors, Stromsberg and Harg, together with several Swedish Lancashire irons; by 1872 they included Gimo iron on their list. In 1873 they supplied Daniel Doncaster and Company with 50 tons of Gimo iron, 100 tons of Stromsberg iron and 1,525 tons of lower grade Swedish Lancashire iron. From 1874 to 1879 Jessops were providing the Sheffield market with Leufsta, Forsmark, Osterby and Gysinge irons whilst Firths sold Gimo, Harg, Stromsberg and Soderfors. By 1880 it seems that the Leufsta iron had also been transferred to the Firth list.

These arrangements appear to have come to an end shortly after this date since, from 1881 onwards, with few exceptions, Daniel Doncaster and Company purchased their Dannemora irons direct from Sweden. In the main, they dealt with the firm of N.M. Hoglund, who supplied Gimo and Leufsta irons together with those from Stromsberg and Gysinge. They also purchased Osterby iron from Carl Setterwall, and the same exporter supplied Forsmark iron. This was available at £19 per ton in 1881 and at £18 per ton in 1886, Leufsta and Osterby irons costing £1 more per ton in each case. In 1891 Bury and Company, another old-established Sheffield steelmaker, had a contract for Forsmark iron at £17 12s. 6d. and was supplying W. Hunt and Company (the famous edge tool makers at the Brades Works near Birmingham, who figured in the counterfeit action brought by Baron de Geer) with blister steel from Forsmark iron at £21 10s. 0d. per ton. In 1893 it was reported that the firm of Nash, again in the Birmingham area, was being supplied by Jessops with blister steel from both Forsmark and Leufsta irons for the making of scythes. The Doncaster purchases in 1895 took on a different pattern, with direct dealing with the Swedish forges. Their contracts for the year were 300 tons direct from Gysinge at £14 15s. 0d., 200 tons direct from Osterby at £20, 750 tons of Leufsta iron from Baron de Geer at £19 10s. 0d. and 200 tons of Gimo iron from Reuterskiold at £20 per ton. The next year, however, dealings were resumed with Setterwall and Hoglund and this pattern persisted well into the twentieth century.

This is, unfortunately a very patchy survey, due to loss of most of the necessary records but it does serve to show that a vital line of communication was from the Swedish ports to Hull, whence the iron was distributed to the Sheffield steelmakers over a period of well over two hundred years and during which the quality of Sheffield steel, which earned it a worldwide reputation, depended very significantly on the availability of high-quality raw materials from Scandinavia, especially the Walloon iron from the Dannemora region.

Note

This paper was commissioned by Vattenfall, the organization responsible for the operation of the Atomic Energy Power Station at Forsmark, which happens to be the location of one of the 'Bruks' or small townships in the Dannemora region of Sweden, north-east of Uppsala, where bar iron was produced by the old Walloon process from the extremely pure iron ore from the Dannemora mines. This iron was highly prized by the steelmakers and, indeed, much of it came to Sheffield. The paper was translated into Swedish and provided the chapter entitled 'Svenskt Jarn och Sheffieldstal' in the publication *Forsmark och Vallonjarnet* (Forsmark and Walloon Iron) sponsored by Vattenfall. This publication of the original English version has been welcomed by Dr Jonas Norrby of Vattenfall, who was responsible for the Swedish publication.

References

1. C. Fox, *A Find of the Early Iron Age from Llyn Cerrig Bach, Anglesey*, Cardiff, 1946, pp. 74–76.

2. D.W. Crossley, *Sidney Ironworks Accounts, 1541–1573*, London, 1975, pp. 237–244.

3. K.C. Barraclough, *Origins of the British Steel Industry*, Information Leaflet No. 7, Sheffield City Museums.

4. L. Ercker, Prague, 1574, translated by A.G. Sisco and C.S. Smith as *A Treatise on Ores and Assaying*, Chicago, 1951, pp. 286-289.

5. F.M. Ress, 'Zur Fruhgeschichte der Zementstahls Herstellung', *Stahl und Eisen*, Vol. 75, No. 15, 1955, pp. 978-982.

6. Patent Roll, 12 James I, Part I, No. 15.

7. R. Plot, *Natural History of Staffordshire*, Oxford, 1686, pp. 374-375.

8. R. Angerstein, *Resa genom England 1753-1755*. Manuscript in Jernkontorets Bibliotek.

9. C. Sahlin, 'Brannstalstillverkningen', Chapter IV, 'Svenskt Stahl', *Med Hammare och Fackla*, III, 1931, pp. 71-103.

10. J. Hunter, in *The History and Topography of the Parish of Sheffield*, (ed. A. Gatty), Sheffield, 1869, p. 214.

11. K.L. Hoglund, 'Brannstalstillverkningen vid Osterby Bruk', *Fagersta Forum*, No. 3, 1951, pp. 11-15.

12. C. Sahlin, 'De Svenska Degelstalsverken', *Med Hammare och Fackla*, IV, 1932. The drawing of the Ersta melting shop is inserted between pp. 42 and 43.

13. J. Hunter, loc. cit., pp. 214-216.

14. K.C. Barraclough, *Crucible Steel Manufacture*, Information Leaflet No. 8, Sheffield City Museums.

15. Rhys Jenkins, 'Notes on the early history of steelmaking in England', *Trans. Newcomen Society*, Vol. 3 (1922-23), p. 27.

16. R. Plot, *The Natural History of Staffordshire*, Oxford, 1686, p. 374.

17. Crowley Council Minute Book, Minutes 39 (4 November 1701) and 41 (11 December 1701). Manuscript in Northumberland Record Office, Gosforth.

18. Staveley Ironworks Records, Reference SIR 1 to SIR 11, Sheffield City Libraries Archives.

19. H. Kalmeter, *Dagbok ofver en 1718-1726 Foretagen Resa*, Vol. i, folios 349-350. Manuscript in Kungliga Bibliotek, Stockholm.

20. H. Kalmeter, *Relationer om de Engelska Bergverken*, 1725, folios 111-116. Manuscript in Riksarkivet, Stockholm.

21. S. Schroderstierna, *Dagbok Rorande Handel . . . 1748-1751*, folios 163 and 200-201. Manuscript in Kungliga Bibliotek, Stockholm.

22. House of Commons Journal, 1737, pp. 853-854.

23. G. Broling, *Anteckningar under en Resa i England Aren 1797, 1798 och 1799*, Vol. 2, Stockholm, 1812, pp. 148-149 and Vol. 3, Stockholm, 1817, p. 51.

24. R. Angerstein, *Resa genom England, 1753-1755*, Vol. 2, folios 190-191. Manuscript in Jernkontorets Bibliotek, Stockholm.

25. Company of Cutlers in Hallamshire, Archive Records 47 and 48, held at the Cutlers' Hall in Sheffield. See also K.C. Barraclough, 'An Eighteenth Century Steel Enterprise', *Bulletin Historical Metallurgy Group*, Vol. 6, No. 2 (1972), pp. 24-30.

26. J.L. Robsahm, *Dagbok over en Resa i England 1761*, folios 9-13 and 20-23. Manuscript in Kungliga Bibliotek, Stockholm.

27. G. Jars, *Voyages Metallurgiques*, Vol. 1, Lyons, 1774, pp. 221-226 and 256-258.

28. B.Q. Andersson, 'Anmarkningar samlade pa Resan i England aren 1766 och 1767', folios 156-161. Manuscript in Jernkontorets Bibliotek, Stockholm.

29. J.H. Hassenfratz, *L'Art de Traiter les Minerais de Fer*, Vol. 4, Paris, 1812, p. 13.

30. Dunn Papers, Document MD 1740-21, Sheffield City Library Archives.

31. Daniel Doncaster Records, Day Book, 1829-1835. Reference BD370, Sheffield City Library Archives.

32. John Fowler, Note Book. Reference WD 374, Sheffield City Library Archives.
33. A. Attman, *Fagerstabrukens Historia*, Vol. II, Uppsala, 1958, pp. 141-145.
34. Daniel Doncaster Stock Book, 1835-1840. Reference BD371, Sheffield City Library Archives.
35. Benjamin Huntsman Stock List, 1843. Reference LD 1618, Sheffield City Library Archives.
36. Daniel Doncaster Records, Iron Purchase Ledger, 1851-1859. Reference Doncaster 50, Sheffield City Library Archives.
37. Tyzack and Sons Purchase Ledger, 1840-1868. Private Collection.
38. Thomas Firth and Sons, Stocktaking Ledger, 1854. Private Collection.
39. F. Le Play, 'Mémoire sur la fabrication de l'acier en Yorkshire', *Annales des Mines*, 4me Série, Tome III, 1843, pp. 602-613.
40. C.V. Heljestrand, 'Om den Finare Smidesmanufakturen i England', *Jernkontorets Annaler*, 1846 (this list is taken from the reference to the original paper in Attmann, *Fagerstabrukens Historia*, Vol. II, p. 157; see Ref. 33).
41. P. von Tunner, *Das Eisenhüttenwesen in Schweden*, Freiberg, 1858 (this list is taken from the reference to the original in C. Sahlin, *Med Hammare och Fackla*, III, p. 143).
42. Daniel Doncaster Records, Iron Purchase Register. Reference DD41, Kelham Island Collection.
43. Percy Papers, Folio Reference 737. Letter dated 15 September 1863. Collection held by Institute of Metals, London.
44. Daniel Doncaster Records, Blister Steel Price Lists. References DD94 and DD95, Kelham Island Collection.
45. 'Stampel-Bok for Stangjerns-smidet vid Svenska Jernbruken', Stockholm, 1845. This particular copy was held by Marsh Brothers, Sheffield, and has recently been transferred to the Sheffield City Library Archives (no Reference Number).
46. D. Mushet, British Patent 2447, 13 November 1800.
47. G. Broling, *Anteckninger under en Resa i England, Aren 1797, 1798 och 1799*, Vol. III, Stockholm, 1816, pp. 69-70.
48. W. Vickers, British Patent 8129, 26 August 1839.
49. S.S. Brittain and Company, Accounts. Reference BD 266, Sheffield City Library Archives.
50. K.C. Barraclough, 'Puddled steel: A forgotten chapter in the history of steelmaking', *Journal Iron and Steel Institute*, October 1971, pp. 785-789.
51. C.W. Siemens, 'On the regenerative gas furnace as applied to the manufacture of cast steel', *Journal Chemical Society*, 1868, pp. 279-310.
52. H. Bessemer, *An Autobiography*, London, 1905, pp. 156-161.
53. E.F. Lange, 'Bessemer, Goransson and Mushet: A Contribution to Technical History', *Manchester Memoirs*, 1913, No. 17. There appears on p. 14 a translation of a letter sent 6 November 1879 to Professor Richard Akerman by Goran Goransson.
54. *London Illustrated News*, 7 January 1860, p. 12. The casting was perfectly sound, stood 5ft 3in high and had a diameter of 6ft 2in at the mouth.
55. S. Pollard, *A History of Labour in Sheffield*, Liverpool, 1959, p. 160.
56. *Sheffield and Rotherham Independent*, 28 April 1874.
57. H. Seebohm, 'On the Manufacture of Cast Steel', printed by Seebohm and Dieckstahl Limited, 1869, for private circulation.
58. F.W. Taylor, U.S. Patent 668,269, 1901. The full history of the development of cutting tools in America is given in F.W. Taylor, 'On the art of cutting metals', *Trans. American Society of Mechanical Engineers*, Vol. 28, 1906, pp. 26-280.
59. Wellmeadow Steel Works, Crucible Steel Charge Book. Private Collection.
60. William Jessop and Sons, Charge Books. Private Collection.

61. P.M. Tyler, 'High speed steel manufacture in Sheffield', *Iron Age*, 10 February 1921, pp. 371-374.
62. Seebohm and Dieckstahl Records. Reference BD97/1, Sheffield City Library Archives.
63. Daniel Doncaster Records, Iron Purchase Register. Reference DD41, Kelham Island Collection.
64. Charles Cammell and Company, Crucible Charge Books. Private Collection.
65. Benjamin Huntsman and Company, Purchase Ledger, 1924-1934. Reference LD 1618, Sheffield City Library Archives.
66. A. Attman, *Fagerstabrukens Historia*, Vol. II, Uppsala, 1958, p. 414.
67. Daniel Doncaster Records, Converting Ledger. Reference DD213, Kelham Island Collection.
68. Staveley Ironworks Records, Reference SIR I to SIR 11, Sheffield City Libraries Archives.
69. See References 30 to 32 and 34 to 38 inclusive relating to the section on 'The Use of Swedish Iron in the First Half of the Nineteenth Century'.
70. Port of Hull Customs Bills of Entry. These commence in 1832 and itemize all imports subject to customs duty. Those for 1832 and for 1837-1852 studied here are available in the Local History Library in Hull; they provide weekly summaries of all imports subject to duty entering the port of Hull. They are valuable in that they itemize bar iron imports under the headings of importer, quantity and port of loading.
71. C. Sahlin, 'Svenskt Stal', *Med Hammare och Fackla*, III, 1931, p. 136.
72. J. Bellamy, 'Some aspects of the economy of Hull in the nineteenth century', unpublished thesis, Hull University, pp. 189-193. Dr Joyce Bellamy also kindly provided additional information from her manuscript notes.
73. Tyzack and Sons Purchase Ledger, folio 80. Private Collection.
74. L.E. Gruner and C. Lan, *L'Etat Présent de la Metallurgie du Fer en Angleterre*, Paris, 1862, pp. 789-790.
75. The Hull Customs records for the years 1853 to 1857 are stated to be normally available in the Guildhall at Hull. At the time of the survey, however, they had been transferred to the Records Office at Grimsby for renovation and binding. The author was permitted to consult them there but found that they were very incomplete; only the pages for May and June 1853, June to December 1854, January to April, July, September and October 1855 and July to December 1857 were available.
76. H.E. Ahrenberg and J.E. Ekman, *Stampelbok for Jernverken i Sverige*, Gothenburg, 1878.
77. Tyzack and Sons Purchase Ledger, folios 95, 102, 121, 133 and 152.
78. Daniel Doncaster Records, Contract Ledgers. References DD63 and DD64, Kelham Island Collection.
79. Daniel Doncaster Records, Miscellaneous Papers. Reference DD41, Kelham Island Collection.
80. Daniel Doncaster Records, Contract Ledger. Reference DD65, Kelham Island Collection.
81. The Hull Customs records for the period from 1858 to 1887 are available in the Local History Library at Hull in more or less complete form. They are now in the form of daily sheets bound in volumes covering each year. Only the 1870 volume was found to be incomplete; the 1869 and 1873 volumes were therefore examined, rather than the 1870 and 1874 volumes as originally planned.
82. A. Attman, *Fagerstabrukens Historia*, Vol. II, Uppsala, 1958, pp. 245-247, p. 452.
83. Daniel Doncaster Records, Contract Ledgers. References DD66 to DD75, Kelham Island Collection.

Intellectual Dependency and the Sources of Invention

Britain and the Australian Technological System in the Nineteenth Century

IAN INKSTER

INTRODUCTION—COMMANDING THEMES

Following the categories recently suggested by Thomas P. Hughes, we might postulate that complex *technological systems* contain four principal components, A-D.[1]

A. Physical technologies—from the Watt engine to the turbogenerator.

B. Organizational technologies—from management procedures to an array of factory layouts.

C: Scientific institutions—from research, through information specification and dispersal, to basic training.

D: Legislative systems—from patent laws to safety regulations.

Such components are highly interactive within any developing system, but I here suggest that *technological progress* is primarily measured through changes at both A and B above. That is, technological progress occurs whenever efficiency (productivity) improvements arise from a series of inventive, innovative and diffusion activities.[2] Engineering changes which do not result in increases in economic efficiency are of less account here, although we acknowledge that the widening of the range of available techniques at time T may well determine both the likelihood and the direction of true technological progress at time $T+i$.

There is a long tradition which explores the causal relations between developments at C and D and effects at A and B. While it is fairly clear that in modern high-technology fields, changes at either or both of C and D may generate decisive technological progress, including the construction of entirely new industries, this sequence is not so observable in a pre-1914 setting. Although nations such as Germany and the USA, and belatedly Japan, created a series of officially directed institutional innovations which were specifically designed to generate technological progress, the relationship between scientific advance and an increase in industrial efficiency in older industrial nations is and was far more problematic.[3]

Given that the Australian colonies were a strategic fraction of the British imperial system, a first theme emerges. Was there a significant relationship between the Australian scientific enterprise and the Australian technological system during the years before 1914? Did technological progress in agriculture, mining or industry result from the diffused application of old or new science? Did the relationship between science and technology in Australia follow the so-called British model?

But, secondly, how far did technological progress in Australia in fact *depend* on British inventive effort? That is, were the advances in Britain at level A, above, significant explanations for the state of the Australian technological system as it was in 1850, 1900 or 1914?

Third, did the intrusion of a foreign (presumably British) technological system retard the development of an innovative Australian technological system? If not, from where did the latter draw its resources—from the scientific culture of Britain (an intruded C above), from its own scientific enterprise (C above), from a purposive, progressive official programme of legislative stimulation (D above) or from elsewhere?

By utilizing data on the Australian patent system from the 1850s—a subsystem of component D above—we can begin to address such questions, for this is highly representative evidence. Unfortunately, our ability to recognize answers depends on commanding assumptions. For brevity we will agree with much of the modern economics literature and recognize or assume that machine technologies (A above) may be very simple and at times impossible to divorce from structures at B above, e.g. the Lombe silk mills of the early eighteenth century.[4] Endowments are given, resources are not—technological change converts the former into the latter, e.g. 'low'-grade ores, gold tailings. Technological change has been commonly unplanned, with spin-offs leading to entirely new industrial processes or sectors, e.g. agricultural chemistry's dependence on basic slag from the Gilchrist-Thomas furnaces. The gap between *invention* and *innovation* is filled by *development*, and the forces giving rise to each of these three processes are not common to them all.[5] Prior to 1914 patents tend to capture invention/development, but at times are a statement of the existence of an innovation. Lastly, a series or spurt of small improvements within an industry or enterprise may be more fundamental to its advance than spectacular, Schumpeterian-type innovations. Incremental change is necessary in order to reduce the establishment or operating costs of a spectacular new technology, e.g. in our period, modifications in the shape and size of furnaces and boilers, improved lubrication, more appropriate metals and alloys in construction, etc.[6] More importantly, best technique may only spread within an industry, between industries or between nations if it is associated with an institutional hinterland of adaptation, diffusion and transfer, most of which activities are incremental in nature, and many of which belong to components B and C above.

THE AUSTRALIAN CONTEXT

Intellectual Dependency

It is less than profound to claim that the Australian scientific enterprise was 'dependent' upon links with Britain and British institutions. It is obvious that the predominance of British mores and institutions, British nationality, a small population base and high level of urbanization, and large distances between population centres in Australia, would yield a dependent scientific enterprise. Together with the drawbacks of the so-called 'convict background' and the disproportionate importance of imperial institutions in governance, such basic features ensured that for some time Australian science would be British science abroad.[7] Australian savants collected evidence, British scientists analysed it. Australia reported findings, Englishmen published them. From mid-century science emerged in Australian universities and other forums, but it was taught, advocated and disseminated by a small band of individuals trained in Britain (predominantly), who were in determined if not always constant contact with Britain, who worked in expectation of rewards stemming from Britain and with the intention of educating a cohort of young scientists whose higher training would take place in Britain.[8]

In the environment of a small-population, staple economy, Australian scientific intellectuals could not expect growing employment in the industrial manufacturing sector, nor could they rely on large remuneration from government unless they engaged in the developmental tasks of the colonial bureaucracies—explorations of land, sea and sky, transport and communications, urban services, commercial acclimatization of foreign flora and fauna, or improvement in the production and transportation of the principal staples, especially wool and gold.

No one has yet tried to quantify the emergence of an independent, Australian scientific enterprise, one which could act as an essential component of an Australian technological system (see above). Although there are several signposts which indicate that through most of our period Australian science remained, in Rod Home's term, 'paradigmatically colonial', there are no measurements of the distance between colonial status and scientific independence.[10] It is fairly certain that no single person, event or institution could traverse the space. The path of transition may be found in answers to such questions as: To what extent did Australian research programmes follow from or replicate those of British science, German science, etc.? What proportion of the publications of Australian scientists were first/eventually published in Britain/elsewhere? How many published collaborations between Australian and British scientists involved a collector–analyst relationship, how many represented an equality of status and effort? At what point could Australian scientists publish work in international journals which was not dependent upon features of locale but which resulted from research that, other things being equal, might have been undertaken in Europe or America?

But even a highly dependent scientific enterprise is not necessarily associated with a dependent technological system, e.g. the USA in the late nineteenth century. A loose association between a dependent scientific enterprise and a developing technological system would only become a causal nexus if progress in the emergent technological system was overwhelmingly the result of contemporaneous scientific applications (see above) *and* if the subsequently transferred techniques required no substantial adaptation to local environmental, economic or institutional conditions. If an increasingly complex technological system—that of Australia in the nineteenth century—is *not* closely integrated into any developed scientific enterprise, and if transferred technique requires frequent and often substantial adaptation, then an independent technological system and indigenous technological progress may feature within the framework of an otherwise 'colonial' relationship. The following sections attempt to show that a viable and independent technological system did develop in the years approximately 1850-1914.

Technological Dependency 1851-1914

Between 1848 and 1918 some 87,000 patent applications were lodged in the Australian colonies. Figure 1 graphs the general trend for the seventy-year period, showing a close relationship with general economic trends, e.g. with the falling away in patenting during the depressions of the 1890s and 1930s. How far was the Australian inventive effort—as indicated in patent applications—dependent upon transfers from Britain and elsewhere? Table 1 below summarizes the occupational and locational characteristics of all 1,851 patentee/inventions applied for through the Victoria patent office in selected years between 1867 and 1896.[11]

Overall, by far the largest single group of patentees were the engineers, 43% of whom were located outside of Victoria. The other large groups of patentees were more colony-based, e.g. 33% of business patentees and 18% of skilled tradesmen patentees were located outside the colony. Engineers were the major carriers of technique into the colony. The table fails to capture trends within the period. The rise of the engineers and businessmen

Table 1 Patent applications in Victoria, selected years 1867-1896

Occupational category	Melbourne	Elsewhere Victoria	All other	Total
Engineers	187	140	244	571
Skilled tradesmen	169	78	54	301
Miners/labourers	7	31	1	39
Merchants	66	5	26	97
Business/capitalists	178	62	118	358
Gentlemen	66	22	47	135
Professionals	37	8	40	85
Others	31	50	38	119
Not known	43	14	89	146
TOTAL	784	410	657	1,851
All partnerships	217	106	202	525

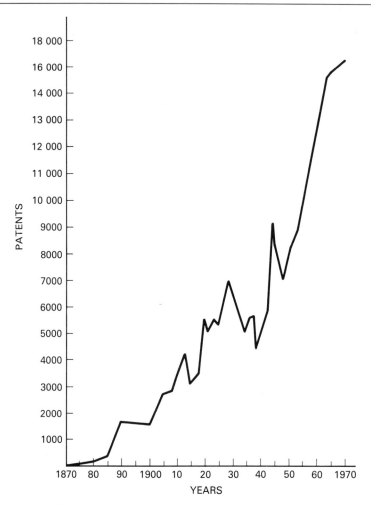

Figure 1 Graphical representations of the number of Australian patent applications lodged between 1870 and 1970. (NB. Years prior to 1904 are based on totals of New South Wales and Victoria.)

is pronounced; of all engineer patentees, 38% patented in 1895–96; this is true of only 29% of all skilled tradesmen patentees and 21% of all gentlemen patentees. The relative importance of engineers amongst the non-Victoria patentees grew throughout these years, moving from 15% in 1867–69 to 33% in 1873, to over 40% during the 1880s and 1890s. The importance of the non-colonial elements overall was increasing throughout the period: of 657 cases, 54% belonged to 1895–96 alone. Partnerships were very common, and 38% of these involved individuals patenting from outside Victoria.

Foreign patentee cases (non-Australasian located) numbered 371, representing 20% of all Victoria patent applications. This is not a large proportion historically (see below). British patenting represented 10% of all Victoria patenting in these years, the USA represented 5%. The manner in which British and American patenting was distributed in terms of both patent type and occupations of patentees is summarized in Table 2. Figures are percentages of each block, with the British listed first; for instance, in block AA, of 35 foreign engineers who applied for gold mining/processing patents in Victoria, 46% were British and 29% were American. It must be remembered that, together, British and American patenting represented 75% of all foreign patenting.

Although British lodgements were twice as frequent as American, the British patenting was less focused on Australia's specific technological needs. Of all British patenting only 23% was in the combined area of food processing and refrigeration, telegraphy and electrical applications, and gold mining; 54% of American patenting was in these fields, i.e. in terms of major developments, America and Britain were about equal at this time in absolute terms.

British lodgements in Victoria were important during the 1860s and early 1870s. These included one by the Kentish Town engineer, J.D. Brunton, for improvements in sinking shafts and pits; the patent of John Hunt of Cornwall (newly arrived in Geelong) for washing and separating gold and other ores; that of the London partners, the assayer W.T. Rickard and the mining engineer W.C. Paul, for gold amalgamating; and that of the partnership between the Danish scientist August Bock and two Scottish manufacturers for the commercial production of stearic and oleic acids by a steaming and chemical process.[12]

During the early 1870s the position of Britain was challenged by a series of applications flowing from Californian-based engineers and manufacturers. Here invention was clearly focused on quartz crushing, amalgamating, concentrating and mine pumping; food processing and refrigeration; sheep

Table 2 British and US patentees as percentages of all foreign patentees in Victoria, selected years, 1867–1896

Patent category	Occupational categories					
	Engineers	Manuf.	Skills	Gents.	Profs.	Other
	A	B	C	D	E	F
A. Gold mining, etc.	46/29	50/50	0/0	66/33	0/0	0/0
B. Other mining	62/25	80/20	0/100	100/0	50/50	80/0
C. Food process/refrig.	23/62	25/42	0/100	0/0	0/0	43/28
D. Agricultural machinery	66/33	75/25	0/0	0/100	0/0	50/50
E. Wool	100/0	0/50	0/0	0/100	0/0	0/0
F. Transport	57/24	17/67	50/0	50/25	0/0	100/0
G. Telegraphy/electricity	24/65	0/0	0/0	66/33	0/100	100/0
H. Other	71/20	48/43	20/40	69/12	57/14	58/17
Total no. of foreign patents	177	76	9	30	17	62

Note: The first figure of each pair is the British percentage, the second the US one.

shearing and wool processing; sewing machines and typewriters.[13] British patenting contributed new rock drills and drilling equipment, but *corporate* patenting was led by the USA and Germany, e.g. the Cologne firm of Kreb Bros. & Co. and Gottlieb Daimler.[14] By the early 1880s the patenting activity of Thomas Alva Edison and George Westinghouse dominated the new fields; in 1881 Edison lodged 16 patents in Victoria. The German presence was represented by Frederick Siemens and Nicholas A. Otto.[15] Incremental patenting had increasingly replaced single lodgements, and corporate patenting was beginning to emerge. Europeans had become increasingly involved in the gold economy, e.g. the patents of Paul G.L.C. Designole for the electrochemical amalgamation of copper ores and for gold separation, and of Louis Thenot for the application of quicksilver to extraction of gold from quartz, sands and tailings.[16]

By the 1890s the environment was a competitive one. British corporate patenting had grown in importance, witness the lodgements of Goodyear, the Floating Metal Co. Ltd. of Boston House, London, the Fleuss Pneumatic Tyre Syndicate Ltd., or the London-based Australian Gold Recovery Co. Ltd. In the field of gold metallurgy there were at least two important British lodgements during 1895—the Sulman-Teed patent, and that for cyanide production of the Scotsman James B. Readman, D.Sc.[17] But American patenting now boasted a strong corporate element, and major thrusts were into the new machine areas, e.g. sewing and typing machinery. In particular, innovation in the general economy was now open to a host of

Table 3 New South Wales patent applications 1855–1887
(first eighteen of 100 categories, 1,196 patents)

Patent category	NSW	Elsewhere Australia	New Zealand	UK	Other	Total
Agriculture	35	51	14	24	16	140
Brewing and distilling	13	11	1	27	3	55
Building	43	16	6	29	15	109
Textiles	7	11	4	11	12	45
Drains/sewage	12	10	1	13	3	39
Elect./magnet/applics.	2	4	1	46	62	115
Engines	18	16	3	42	28	107
Excavation/dredging	18	12	1	1	3	35
Explosives/firearms	1	9	0	7	12	29
Prep./preserv. food	30	27	2	37	30	126
Fuel	19	9	2	12	7	49
Coverings (clothing)	5	9	0	6	4	24
Livestock	13	15	1	4	4	37
Illumination	27	9	3	28	13	80
Indicating mechanisms	10	9	2	9	5	35
Leather/hides, etc.	5	9	0	6	7	27
Marine/submarine	16	8	3	9	3	39
Minerals and mining	48	24	2	14	17	105
TOTAL	322	259	46	325	244	1,196

inventors, e.g. the patents of the engineers Louis Pelatan of France and Fabrizio Clerici of Italy for the electrical separation of the precious metals from their ores and for desulphurization furnaces, and the similar effort of the Berlin-based chemist, Dr Albrecht Schmidt.[18]

Victoria was not peculiar. The New South Wales (NSW) patent data show the same competitive environment and little evidence of a specifically 'colonial' technological relationship. Table 3 provides details of 1,199 patent/patentee applications in the colony of NSW for the first eighteen of one hundred industrial categories for all years between 1855 and 1887.[19]

For these categories, foreign patenting represented 48% of all applications in the colony of NSW. Although British patenting was particularly important in such areas as food preservation, telegraphy and electrical equipment, it had no monopoly. Indeed, in the latter field American patents[50] just outmatched British.[46] In such key areas as building, mining and mineralogy Britain did not loom large. Again, although British corporate patenting was of increasing importance, so too was that of such American companies as Ewart Manufacturing Ltd. and the Llewellyn Steam Combustion Manufacturing Co.

Table 4 is more persuasive, for it clearly illustrates that total foreign involvement in the Australian technology system—as measured by the patent data—did not increase between 1886 and 1918, that British (including British dependencies') activity faced an almost secular decline from its high point of the 1880s, and that American patenting in Australia had matched that of Britain by the turn of the century. The period from the end of the long boom, the depression of the 1890s, the recovery and the First World War years was not one of British technological colonialism. Moreover, British lodgements took place in an increasingly competitive environment: to America were added the rapidly industrializing nations of Germany, Sweden, Denmark and others, as well as the closer involvement of the commercial economy of France. Figure 2 illustrates how foreign lodgements overall did not lead or lag, but on the whole coincided with the pattern of Australasian lodgements.

Such points are brought out more clearly in Tables 5 and 6, which are based on a detailed breakdown of foreign patenting in the year 1906, by which time the new Commonwealth system of patenting was fully operational and American patenting was equal to British. During 1906 there were 2,743 patent applications in Australia. Of the 1,181 individual foreign patentees in 1906, 76% were resident in either Britain or the USA, with the latter reaching a slight predominance. More significantly, Table 5 shows that USA patenting was in advance of Britain in terms of corporate and business activity. British patenting in Australia still relied greatly on the individual engineer or tradesman (to 59% of the UK total).

The first row of Table 6 shows the number of all individuals involved in foreign patent partnerships, a total of 667. Compared to both Britain and 'all other', American involvement in the Australian technological system stands out as a joint venture between invention and entrepreneurship, skills and capital; only 57% of American patentees were active inventors, the rest being assignees, nominees and communicators who had at least partially

Table 4 Foreign activity in the Australian patent system 1886–1918

Nation	Applications in New South Wales						Applications in Australian Commonwealth					Rank 1886–1903	Rank 1904–1918	Rank 1886–1918
	1886–88	1889–91	1892–94	1895–97	1898–1900	1901–03	1904–06	1907–09	1910–12	1913–15	1916–18			
Australasia	1,044	1,701	1,509	1,775	1,701	2,028	5,566	6,558	8,225	8,111	7,513	1	1	1
Britain	270	413	289	398	418	406	810	846	1,068	910	825	2	3	2
British Dependencies	10	12	12	33	39	68	166	197	145	152	79	5	5	5
USA	96	162	160	234	332	408	894	904	1,121	1,024	921	3	2	3
Germany	26	21	32	38	39	48	138	205	292	179	17	4	4	4
France	20	17	9	32	45	28	97	102	124	77	55	6	6	6
Sweden	8	6	5	10	7	29	57	58	56	78	96	7	7	7
Denmark	1	1	2	5	23	8	21	34	24	13	20	8	9	9
Austria	2	10	3	3	6	8	26	20	19	12	0	9	14	12
Belgium	6	4	1	8	9	3	14	33	27	24	3	10	11	10
Russia	3	2	2	2	4	3	4	17	15	3	0	11	16	16
Italy	0	0	1	6	2	5	26	25	15	23	22	12	10	11
Switzerland	2	0	1	1	1	7	20	21	20	8	13	13	13	13
Norway	0	0	1	1	1	4	8	6	14	12	24	14	15	15
Latin and C. America	2	2	0	3	0	0	0	3	0	10	11	14	17	17
Netherlands	0	0	0	0	5	1	4	18	6	24	33	16	12	14
Spain and Portugal	0	2	2	1	0	1	2	2	4	3	0	16	19	19
Japan	0	0	1	0	0	0	0	2	3	6	11	19	18	18
Elsewhere	0	3	0	0	1	0	13	21	27	69	55	18	8	8
Hungary	0	0	0	0	0	0	0	0	2	9	0	20	20	20
TOTAL	1,490	2,356	2,030	2,550	2,633	3,055	7,866	9,072	11,207	10,747	9,698	–	–	–
% Foreign	30	28	26	30	35	34	29	28	27	25	23	–	–	–
% UK and dependencies	19	18	15	17	17	15	12	11	9	11	9	–	–	–
% USA	6	7	8	9	13	13	11	10	10	10	9	–	–	–

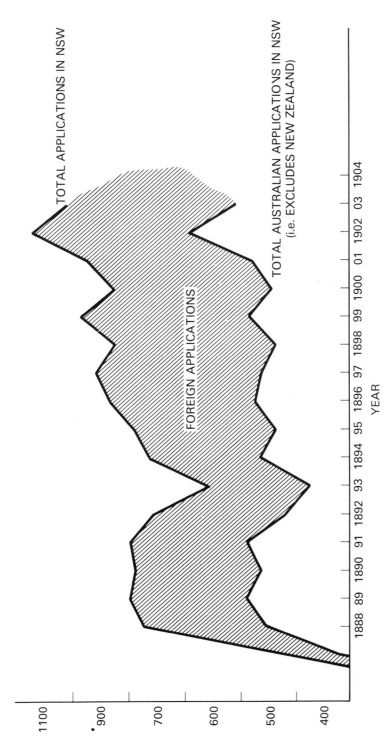

Figure 2 New South Wales patent applications, the foreign element.

taken over the intellectual property right of the original inventors. Patent agents are excluded throughout.

From whatever location, partnerships were overwhelmingly of individuals *within* the same occupational category, and these were dominated by engineers and small businessmen. Of the remainder, Americans were more prone to partnerships between corporations and others, particularly

Table 5 Foreign patentees in the Australian patent system, 1906

Occupational categories	Britain and dependencies		USA		All other	
	Number	% Active inventors	Number	% Active inventors	Number	% Active inventors
A. Engineers	217	97	180	97	97	97
B. Manufacturing chemists	13	100	11	100	18	90
C. Skilled tradesmen	26	100	30	100	12	100
D. Business/manufacture	39	100	62	77	42	86
E. Merchants	15	93	4	75	15	86
F. Managers and contractors	8	100	2	100	10	80
G. Retail/agents	12	92	12	83	0	–
H. Commercial professions	10	80	3	66	1	100
I. Other professions	7	100	15	80	13	100
J. Corporate patenting	49	26	112	23	54	20
K. All other	39	92	32	65	21	95
TOTAL	435	–	463	–	283	–
% Corporate	11		24		19	

Table 6 Foreign patent partnerships in Australia, 1906

Nature of partnership	Location		
	Britain	USA	All other
Total individuals involved	251	263	153
% of individuals who were active inventors	76%	57%	75%
Number of partnerships			
Tradesmen with engineers	6	2	0
Tradesmen with tradesmen	1	2	1
Tradesmen with gps. D–K (Table 5)	6	4	2
Engineers with gps. D–K (Table 5)	19	13	11
Corporate with engineers	31	53	36
Corporate with tradesmen	0	11	1

engineers. It is the rise of corporate patenting—where the corporation *per se* secures patent rights—which best illustrates the competition amongst foreign interests in the Australian system. Although the total of 'all other' patenting was far less than that of Britain, the absolute number of corporate patents and corporate-engineer partnerships stemming from these nations was greater than the British.

Several of the forty-seven Australian patent attorneys specialized in the processing of foreign applications, and this was the normal mode of penetration for the foreign corporations. For instance, the Melbourne patent agent, George G. Turri, handled, in 1906 alone, important applications from such US corporations as Renstrom Tempered Copper Company (hardening of copper), the Iler Rock Drill Manufacturing Co. (hammer drills), the Vermont Farm Machine Company (centrifugal separators), the Garvin Cyanide Extraction Co. (gold recovery) and the Blaisdell Company (agitators), as well as applications from a wide variety of American and European engineers.

Of course, British patenting remained important in early twentieth-century Australia. Individual efforts could be of significance. British-located engineers and chemists with research degrees from European institutions lodged patents for a variety of inventions, from the explosives of Oswald Silberrad of Kent to the driving mechanisms of Arthur W. Brightmore of Surrey. The most important series of incremental inventions by a British scientific-engineer were those of Ralph W.E. MacIvor for the treatment of ores containing nickel, of sulphide ores containing zinc and of complex ores containing gold, which were obviously based on a substantial individual experimental programme.[20] Metropolitan and provincial engineers lodged a series of applications in key areas—rock drills, gold extraction, systems of electrical distribution (e.g. the application of the electrical engineer Randolph Braun of the Westinghouse Works, Manchester), recycling of waste materials (e.g. the application of Adolf Gutersohn, metallurgical chemist at the Cromer Works in London), and the internal combustion engine.[21] Again, there is much evidence of technical research programmes, particularly in the lodgements of the well-known engineers Henry L. Sulman of London and Charles A. Parsons of the Heaton Works, Newcastle. Both illustrate the value of partnerships and assignments. Sulman applied for no less than ten distinct patents in 1906 through inventive association with engineers and metallurgists and financial association with merchants such as John Ballot of London. These patents were incremental in nature, involving the separation of metalliferous minerals, improvements in ore concentration and treatments of zinc ores.[22] In one instance Sulman was assisted in his application by the Melbourne civil engineer, Charles Hatton. After an involvement in the old colonial patent system, Parsons lodged a series of eleven Commonwealth patents in 1906 in the areas of electrical transmission, turbo compressors, turbine shafts and blades, and electric dynamos.[23] These involved him in inventive partnership with several engineers, including H.L. Short of Surrey, Hewitson Hall and John Turnbull of the Turbinia Works, Northumberland, and Alfred Flint, George Storey and Alexander Law, all of Heaton.

Partnerships between engineers and between engineers and businessmen and corporations also generated a series of British patent applications, and acted as a key entry mode for the expert who would otherwise have found it difficult to secure financial backing in Britain. Thus the engineers G.W. Beynon and G.H. Mackillop joined with the company director Gustave Bonnard to bring out improvements in 'apparatus for pulverizing, crushing, stamping and other operations'.[24] A Linotype operator of London, J.G. Holbourn, joined with the engineer H.A. Longhurst in bringing out a series of colonial and Commonwealth patents on type matrices for Linotype machines, whilst two engineers of the Craven Iron Works at Salford issued a series of similar incremental inventions. From the traditional mining locales of Cornwall came specific expertise: the engineers John and James Holman patented improvements in rock drills, whilst W.C. Stephens registered in partnership with the Camborne Rock Drill Works for a patent concerning adjustable guides for drilling.[26] The Marylebone chemical engineer, Guy de Bechi, required the financial partnership of the London merchant, Reginald W. Rücker, in applications for 'improvements relating to the treatment of complex sulphide ores', whilst the Scottish engineer, David M. Ramsay, brought out patents for improvements in locomotive engines in conjunction with Hugh Reid, managing director of the North British Locomotive Co. Ltd.'s Hydepark Works in Glasgow.[27] The partnership of the engineer, B.H. Bedell, and the merchant, William Griffiths, yielded a series of colonial and Commonwealth patents relating to electrical traction, all of which were mounted as a joint business venture from Hamilton House in Bishopsgate.[28] Within the Castner-Kellner Alkali Co. of Cheshire was spawned the partnership of manager A. Thomas Smith and chemist Harry Baker, which produced an 'improved process for the treatment of sulphide ores'.

British corporate patenting was of growing importance. Amongst this group of applications were several examples of the corporation simply capturing the patent rights of a corporate entrepreneur's own invention. This is true of the applications of Pillatt and Co. Ltd., of Nottingham (furnaces), Marconi Wireless Telegraph Co., the de Forest Wireless Telegraph Syndicate Ltd., the Bifurcated Rivet Co. Ltd., and the Incandescent Heat Co. Ltd. (furnace systems). More commonly, the corporation had bought into the property rights of an independent engineer-inventor. This was the case with such important lodgements as those of the Liverpool firm of Weldite Ltd. (alloys), the I.R. Refractory Ore Syndicate Ltd. (treatment of pyritic ores) and the Western Syndicate Ltd. (railway signalling), which involved four inventor-engineers from Reading, Cardiff and Hanwell. The New Century Engine Co. Ltd. (locomotive engines and generators), Neophone and Machinery Ltd., J. Stone and Co. (air conditioning), the Edison Ore Mining Syndicate Ltd. (magnetic separators) and the Malcolm Fraser Wheel Syndicate Ltd. and several more were all British corporations in which patenting represented a transfer of at least some property rights from the individual inventor to entrepreneurial interests. Amongst these were a few firms who took up the patent rights of foreign inventors—British Westinghouse (American mechanical engineers),

Stewarts and Lloyds Ltd. (a South African engineer, valves) and the Imperial Fibres Syndicate Ltd. (an Australian civil engineer, decorticating fibrous materials).

But whilst British inventive activity was developing a commercial face, it yet relied on the tradition of individual inventive activity based on specific industrial skills. A more commercially oriented invasion of the Australian technological system originated in other nations, particularly the USA, and is illustrated crudely in the higher proportion of corporate patenting and partnerships. It is also shown in the somewhat different character of American corporate patenting in Australia, most of which corporations were acting as the commercial property-right holders of a variety of independently generated inventions.

Within an array of American corporate applications, three types stand out. Amongst the most vigorous patentees were corporations explicitly established in order to exploit an interrelated series of new or existing inventions, home or foreign in origin. The Boston-based United Shoe Machine Co. brought out no less than twelve patents in Australia in 1906, all of which were on assignment from engineers throughout America, Britain and Australia.[29] Other corporations within this group included Westman Process Corp. of New York (iron ore reduction), the International Steam Pump Co. (an assignment with the German inventor C.H. Jaeger, centrifugal and turbine pumps), George Westinghouse Ltd. (a series of colonial and Commonwealth patents concerning pumping machinery, duplex system engines) and Morris International Patent Co. (refrigeration).

A second type of company was the one specializing in new areas of equipment and product supply, such as sewing machines and typewriters. Amongst these were Brunswick Refrigerating Co., the J.P Karns Tunnelling Machine Co., Rhekheim Bros. and Eckstein (waterproofing), the Globe Rotary Engine Co., International Telegraphic Call Co., Mergenthaler Linotype Co., the Electric Boat Company (whose patent series included a number of independent engineers), Decker Electric Manufacturing Co. and the Brown Hoisting Machinery Co. (a variety of electrical equipment in a series of patents).[30]

Finally, several American corporations applied for patents in fields especially appropriate to the exploitation of Australian staples. Such companies included Robins Conveying Belt Co. (conveyance for ores); the Holland Metal Recovery Co.; an important series of eight patents lodged by the New York-based Ingersoll-Rand Co. in the fields of coal cutting machinery, electro-pneumatic channellers, rock and hammer drills and compressors; C.T. Carnahn Manufacturing Co. (mining tools, pneumatic drills); the Iler Rock Drill Co. (hammer drills); Vermont Farm Machine Co.; and the Garvin Cyanide Extraction Co.

These three types loomed far larger in American than in British corporate patenting, but were to be found also in German and Swedish applications. One of the most active corporations of 1906 was the Swedish firm of Aktievolaget Separator, which applied for twelve patents in Australia. These all involved farm and dairying equipment and originated in partnerships with several independent Swedish engineers.[31] As with foreign patentees at large, German corporate patenting included a range of electrical applications, but was better represented by mining and metallurgical invention. The Berlin manufacturers Accumulatoren-Fabrik Aktiengesellschaft patented in the area of lead and silver extraction, whilst Rheinisch-Nassauische Bergwerk of Stolberg patented equipment for ore dressing and processes for the extraction of zinc from its ores. The firm relied on a variety of engineer-inventors, amongst which was the partnership of the academic technologist, Professor Wilhelm Borchers, and the metallurgical engineer, Arthur Graumann.

In essence, although there was a significant amount of foreign involvement in the Australian patent system over most of the years considered here (see Table 4), Australian-originated patenting dominated the field. Australia was open to foreign technological direction, but the nature of Australian dependency on foreign technique does not appear to have been peculiarly 'colonial' in character. Despite cultural and institutional advantages, British inventive efforts had to compete with those of other nations. Furthermore, the USA in particular appears to have exerted an influence on the Australian technological system which was more commercially viable than that of Britain.

THE AUSTRALIAN INVENTORS

Outside the major cities, patentees were distributed between pockets of urbanism based on specific industries or staple exploitation, e.g. the gold towns of Geelong, Broken Hill, Ballarat, Maryborough, Kalgoorlie and so on. As an example, between 1867 and 1869 the town of Ballarat produced fifty-three cases of patentee-inventions, or 17% of the total cases emanating from Victoria in those years. Most of such non-city patents came from engineers, skilled tradesmen and miners—accounting for 66% in this example. The importance of individual inventive effort suggests the significance of skills relating to locale. The Ballarat example bears investigation.

The generalizations which follow are based on a detailed study of the Ballarat patenting, skill and training environ in the years 1850–1918.[32] Ballarat was a gold town in transition from simple alluvial mining to complex quartz mining and ore processing.[33] The working of the more difficult reefs together with the challenge of developments elsewhere in the colony (which periodically pulled resources and people out of the sub-region) promoted disproportionate inventive activity in and around Ballarat. Several points relating to the origins of invention are worth noting in brief. As might be expected, invention was highly locale-specific in the early years. During 1867–69 Ballarat patenting centred on the working of alluvial dirts, processes for separating gold from tailings, the design of pyrites furnaces, puddling and sluice machines, pumping engines, improved buddles and drilling equipment, and amalgamation. From the 1880s the area developed as a centre of foundries and machine making, entailing the production of mining equipment, agricultural machinery and locomotives.[34] Patenting became more wide ranging, with applications for grain stripping machinery, balers and chaff cutters, stump jumping improvements, wire and fencing equipment, drain cutting machinery, refrigeration, electrical signalling, railway equipment and a variety of non-mining-specific machine improvements (furnaces, boilers, lifting jacks); moreover, there was a movement of mining patenting towards ventilation, conveyance and safety. Thirdly, patentees were commercially and socially mobile. Such men as Thomas Moore and Ware Copeland, miners when they patented in the late 1860s, had become assayers and sharebrokers (respectively) by the 1880s. Several patentee engineers became prosperous local ironfounders in subsequent years. The engineer George R. Kerr, who was patenting stampers during the 1870s, was by the early 1880s the proprietor of a company of ironfounders, agricultural implement makers and steam machinists. Others, such as William Henry Shaw, became managers of machine emporiums—in this case the Phoenix Foundry, which produced locomotives and agricultural machinery. Commonly, patent partnerships became the source of business enterprise: two general smiths, James Kelly and John Preston, patentees of improved ploughs in the 1860s, were by the 1880s the joint proprietors of an agricultural machinery firm. Several set up in business only as a short-term measure for exploitation of a particular patent, producing cams for driving shafts, safety devices for mining cages, harvesting machinery, crushing machinery and so on. A large proportion of skilled tradesmen patentees became small businessmen thereafter.

Two measures suggest that Australian-based patenting was central to the overall technological system. First, though in a small way, Australian corporate invention did emerge in the last twenty years of the nineteenth century. Mostly these firms were set up to exploit a particular line of patenting, and they included the Tasmanite Manufacturing Co., the Mallac Patent Electric Lighting Co., Lysaght Bros. and Co. Ltd. (sheet iron and fencing) and the Colonial Sugar Refining Co. (sugar cane diffusion apparatus). Second, and of greater importance, Australian patenting was never a random mix but was the product of surges of invention in particular areas or industries. Although inventions for food preservation are found throughout these years, a surge of refrigeration patenting appeared in 1867-74. A building materials surge of the 1880s involved British patenting but was led by Australian inventors concerned with improvements in Portland and other cements, bitumen, damp-proofing, tiling, brick making, pavement materials and treatment of timbers. On the other hand, the motive power surge of the 1870s was led by the foreign element at large (steam, compound engines, gas engines, etc.), but induced incremental Australian improvements in pistons, cylinders, governors, safety valves and so on, applied to a variety of purposes—card winding, mill bands, etc. Whilst the electronics surge of the 1880s (accumulators, dynamos, electro-magnetic equipment, power transmission systems, insulators, telegraphs, signalling) was certainly led from abroad, the more mundane excavating and dredging inventions of the 1870s were predominantly Australian. It seems that Australian invention was focused on local needs and surged in years of particular demand. This is well illustrated in the mining and

Table 7 Australian mining and metallurgical patents, 1892-98[35]

Residence of patentees	WA	Vict.	NSW	SA	Tas.	Qld.	Two or more Australian colonies	NZ	Total
Victoria	1	18	1	0	1	1	11	1	34
NSW	0	3	10	1	0	0	9	0	23
Other Australia	11	4	2	3	0	3	6	2	31
New Zealand	0	1	3	0	0	0	6	5	15
Britain	7	6	7	1	2	2	23	1	49
USA	1	2	1	1	1	3	6	1	16
Other foreign	0	1	4	0	0	0	8	3	16

Table 8 Commonwealth patenting, 1905-1911

	1905	1906	1907	1908	1909	1910	1911
Manufacturing — total	1,458	1,756	1,883	1,481	1,564	1,946	1,515
Motive power	366	389	447	317	313	355	239
Metal-working	284	313	301	262	224	298	249
Telegraphy	33	35	42	25	44	71	55
Other electrical	130	188	160	102	95	116	75
All other	645	831	933	775	888	1,106	897
Mining — total	459	523	496	312	320	359	223
Machinery	213	210	205	132	137	147	78
Metallurgy/mineral	193	255	228	130	141	169	106
All other	53	58	63	50	42	43	39
Food processing/preserv. — total	231	286	328	250	243	270	226
Refrigeration	29	44	43	29	33	48	38
All other	202	242	285	221	210	222	188
Public works — total	344	479	510	410	459	543	453
Building	149	222	264	182	197	261	229
All other	195	257	246	228	262	282	224
Agriculture — total	313	352	408	245	315	390	336
Machinery	186	209	265	155	208	280	230
All other	127	143	143	90	107	110	106
Transportation — total	425	450	470	466	520	618	521
Railway	244	273	298	271	291	385	310
All other	181	177	172	195	229	233	211
Other miscellaneous	582	631	704	648	710	811	647
TOTAL	3,812	4,477	4,799	3,812	4,131	4,937	3,921

mineralogy patenting of the 1880s to 1900s, i.e. from the end of the long boom, through the resource-scarce depression of the 1890s, to recovery in the 1900s. New South Wales patenting increased in this field throughout the 1870s and was joined as a surge of applications in Victoria and elsewhere during the 1880s. Prior to that time many key advances came directly from abroad. During the 1860s American engineers had lodged a series of patents for gold reduction and amalgamation. By the 1880s they had been added to by French and British patentees, such as Jules Weinrich, Paul G.L.C Designole (chlorides in gold extraction), Bernard Molley (amalgamation), E.W. Parnell and James Simpson (gold from antimony ores), William White (centrifugal amalgamation) and the British cyaniders. But at the same time Australian involvement intensified also, and included the important technical advances of John Hunt (washing and separating), a series by James Cosmo Newbery in partnership with others (extraction from pyrites and antimony ores), and the applications of Henry Herrenschmidt. For the years 1892-98 the residence of Australasian patentees whose specifications were accepted in this field are shown in Table 7. Fairly clearly, foreign, especially British, activity was important, but it was complementary to Australian inventive activity. This was almost certainly also true in areas such as food processing and agricultural machinery.

Table 8 provides a rationalization of the ninety-nine categories of patents applied for in the years 1905-1911, amounting in all to 29,889. A very large number of applications, 11,630 (39%), were in manufacturing. In this sort of exercise 'all other' leaves much to be desired. However, in the key engineering areas of motive-power and metal-working innovations there were no less than 4,357 patent applications. Invention in mining yielded another 2,692 patents (9% of the total). The growing concentration of Australian inventive efforts meant that miscellaneous patenting as such never reached more than 17% of the total. Even if the dubious category of 'all other' in manufacturing is entirely re-allocated to 'other miscellaneous', the total amount of patenting in areas outside those fundamental for Australian economic development amounts to 10,808 or 36% of the total. This still allows the Australian technological system to be characterized as one closely related to the requirements of Australian economic development in the years prior to the First World War.

A final return to the 1906 data provides our last overview of Australian-based patenting. Table 9 breaks down all Australian residents by occupation and region and compares them with patentees located in New Zealand and Britain. Obvious points relate to the very small proportion of non-active patentees involved in Australian-located patenting and its highly urban nature: 43% of all Australians patenting in 1906 were located in either Sydney or Melbourne. Within the large cities business, retail and agency patents were of disproportionate importance. On the other hand, skilled tradesmen were of special importance outside the major cities. Also very clear is the tiny amount of Australian corporate patenting. Engineers and skilled workers predominated overall and were seldom anything other than active inventors. Table 10 breaks down all Australian partnerships, comparing them with those of Britain and the USA.

Table 9 Occupational characteristics of Australian patentees, 1906

Occupational categories	Sydney No.	Sydney % Active inventors	Other NSW No.	Other NSW % Active inventors	Melbourne No.	Melbourne % Active inventors	Other No.	Other % Active inventors	All Australia No.	All Australia % Active inventors	NZ No.	NZ % Active inventors	UK No.	UK % Active inventors
A. Engineers	106	96	38	94	119	97	91	96	477	96	53	96	217	97
B. Manuf. chem.	7	85	0	–	6	83	1	100	16	90	1	100	13	100
C. Skilled trade	59	96	63	98	116	96	157	99	458	98	32	100	26	100
D. Business	51	88	10	90	76	86	42	90	221	92	16	93	39	100
E. Merchants	20	70	6	50	16	62	6	50	64	70	11	100	15	93
F. Managers, etc.	19	78	14	100	26	92	21	100	102	86	9	100	8	100
G. Retail/agent	41	90	18	83	50	90	40	85	209	80	22	86	12	92
H. Comm. profs.	9	88	3	100	9	77	15	100	62	90	4	100	10	80
I. Other profs.	19	78	6	100	13	76	12	92	68	90	6	100	7	100
J. Corporate	2	50	5	20	6	50	1	100	21	38	2	100	49	26
K. All others	46	86	46	98	49	85	29	93	300	94	49	98	39	92
TOTAL	379	–	209	–	486	–	415	–	1,998	–	205	–	435	–
% Corporate	0.5		2		1		0.2		1		0.9		11	

Table 10 Australian and other patent partnerships, 1906

Nature of partnership	Australia	Britain	USA
Total individuals	875	251	263
% Active inventors	85	76	57
Number of partnerships			
Tradesmen with engineers	16	6	2
Tradesmen with tradesmen	46	1	2
Tradesmen with gps. D-K (Table 9)	63	6	4
Engineers with gps. D-K (Table 9)	84	19	13
Corporation with engineer	9	31	53
Corporation with tradesmen	2	0	11

Partnerships clarify matters a little. Given that most patent partnerships anywhere were between engineers or between small businessmen, and given the high proportion of those groups in Australian-based patenting, then the relatively numerous partnerships between Australian-located tradesmen, tradesmen in alliance with higher social groups, and engineers in alliance with higher social groups, shows the individual and skill-based characteristics of the Australian technical system clearly. Across the board the number of individuals in partnerships as a percentage of total patentees was approximately the same for Australian, British and USA lodgements, but the Australian partnerships were more based on mutual active invention and did not, in the main, involve large-scale commercial interests.

SUMMARY AND CONCLUSIONS

In the years prior to 1914 the Australian technological system was capable of generating changes of technique and technological progress. Using patents as a measure once more, Table 11 suggests that by the early twentieth century per capita inventive activity in Australia was on a par with that of other nations.[36]

From the 1880s there is little evidence of an especially 'colonial' technological dependency relationship between the Australian colonies/Commonwealth and Britain. Throughout the period foreign inventors, partnerships

Table 11 Patenting in selected nations, 1904-1916: number of applications (applications per 10,000 of population given in parentheses)

Nation	1904	1912	1916
Australia	2,563 (6.4)	3,436 (6.9)	3,543 (7.1)
Canada	5,793 (10.4)	8,681 (11.2)	6,812 (8.1)
Britain	29,655 (7.0)	30,102 (6.5)	18,602 (4.3)
New Zealand	1,491 (17.4)	1,769 (16.3)	1,388 (12.5)
USA	50,213 (6.6)	70,976 (7.4)	68,075 (6.7)
France	–	15,735 (3.9)	15,967 (4.0)
Germany	–	44,929 (6.6)	49,532 (7.3)

and corporations competed for a place in the Australian technological system, and, in doing so, created what Gustav Ranis has labelled a 'technological shelf', an array of techniques from which Australian producers could finally choose.[37] At the same time, foreign intrusion in the Australian technological system—as measured by patent data—was not overwhelming. Within this framework British involvement was giving way to that of other nations, especially the USA.

No nation's technological system is closed, though at times legal and other measures may be utilized in an effort to reduce foreign interference. Thus outright discrimination against foreigners in the American patent system meant that of over 7,000 patents lodged there between 1836 and 1849 only 3% were foreign in origin. Even during 1870–1900, when an average of nearly 19,000 patents were lodged in the USA annually, only 9% were foreign. But US technical capability was boosted primarily by direct inflows of 'human capital' and by selective purchase of European technique, or informal transfer.[39] In a much smaller scale this is almost certainly what was happening in the Australian system too, but such modes of technology transfer are not formally 'colonial' in nature, i.e. they do not result primarily from an institutional net designed to promote or maintain a colonial relationship.

We may compare Australian dependency as measured by patents with that of other nations. Canada is seemingly a most apt example, and Table 12 shows the successful registration figures for selected years between 1892 and 1905.[40] The Canadian patent system owed everything to foreign transfers, and was particularly subject to the economic proximity of industrial America. With the exception of 1897–98, the period 1892–1905 was associated with a strong inflow of US capital into Canada, into the nickel industry of northern Ontario, the mills of the Gaspe Basin, asbestos, mica, gold, textiles, etc.[41] A high proportion of the new industrial and mining concerns of Canada were owned by or dependent upon US capital.[42]

A comparison between Australia and New Zealand shows far more common ground. As Table 13 illustrates, foreign involvement approximated to that of Australia, and British patenting represented some 10% of all New Zealand applications.[43]

In summary, a comparison of these three staple regions shows that British colonialism led to far less dependency on one foreign source in the case of Australia and New Zealand than did proximate US capital in the case of Canada. What of developing industrial nations? Germany and Japan are

Table 12 Canadian patentees (successful registration), selected years 1892–1905

	1892	1894	1899	1901	1903	1905
Total patentees (reg.) Canada	3,417	2,756	3,151	4,586	5,714	6,111
Total Canadian patentees	671	661	601	744	794	888
% foreign patentees	80%	76%	81%	84%	86%	86%
% USA patentees (/total)	65%	62%	65%	71%	74%	73%

Table 13 Patent applications in New Zealand, 1884–93

	NZ patents	Foreign patents	Total	% Foreign
1884–86	880	304	1,184	25
1887–89	1,351	636	1,987	32
1891–93	1,384	756	2,140	35

prime examples of large nations whose industrial revolutions were at key points dependent on technology transfers. In the period 1877–1894, 28% of all applications to the German patent office were by nationals of other countries.[44] Japan created major obstacles against foreign patentees; but even here, of 18,616 applications between 1896 and 1904, 10% were of Western origin, and half of these were from American nationals.[45] The most distant technological culture was by no means closed. It might be noted that even the 'Workshop of the World' faced a barrage of foreign applications during 1867–69; the major city from which patentees applied for British patents was certainly London (2,480), but the second was not Manchester but Paris (966), the third not Birmingham but New York (393).

So the Australian technological system was not in fact starkly open to foreign influence when compared to that of other nations at that time. The level of exposure was possibly optimal for economic development in the more general sense. Foreign patents may have been designed to block indigenous invention, but they almost certainly resulted—as the Australian case suggests—in a vigorous culture of incremental and circumnavigational (i.e. avoidance) inventive activity from within the indigenous technological system.[46]

Australia until 1914 produced a technological system capable of sustained incremental invention. In the absence of a multitude of small and medium-sized enterprises based on technique improvements and their commercialization the Australian economy would have been far more exposed to foreign ownership and control than was the case. British capital inflow may have been of overwhelming importance to the growth of the economy in certain decades, yet even at such times this did not dampen the development of a complex of skills, business enterprise and innovative institutions which was genuinely Australian. There is little doubt that the principal function of the Australian technological system was to permit the introduction of foreign technique into industry, mining and elsewhere in a manner reasonably appropriate to the needs of the staple economy. In addition, a viable skill system was required in order to select efficiently from the array of foreign techniques available, many of which were inappropriate without this adaptive process.

By 1958 the percentage of patent applications lodged in Australasia by residents was only 41%—a figure nearer that of India today than Australia prior to 1914—and to the late 1960s the growth of Australian patenting was a direct product of accelerating foreign involvement, particularly the fast penetration of American corporate rule.[47] While there may be institutional and even intellectual arguments to explain the reversal, a likely explanation for increased dependency might involve the changing nature of commanding technological imperatives: best technique is now more capital- and skill-intensive, requires a large scale of operation, and therefore market size, and is often more environmentally specific than in the case of pre-1914 technique.

There is as yet little reason to be convinced that increased dependency has impacted negatively upon Australian economic growth or upon the Australian technological system. Blumenthal has argued that Australia joined with France and Japan as a nation which has evidenced a positive correlation between the importing of advanced technologies and the development of internal research and development capability.[48] During the years which we have examined here, Maddison has shown that Australia ranked remarkably high amongst nations in terms of overall productivity growth, admittedly a dubious concept.[49] Technology transfer in both periods, though in different ways, seems to have allowed growth in the technological system. In the right conditions, imports of technology may act as catalysts of increased R & D activity. Lamberton's statement of nearly twenty years ago strikes the relevant chord:

> The derivative nature of much of the Australian research effort is merely one aspect of the derivative nature of the Australian economic and cultural pattern. This does not imply that a domestic component is not discernible, nor that the component is in any way inferior.[50]

We might now add that in the years approximately 1850–1914, the Australian technological system generated techniques in complementary association with foreign transfers, and that the latter were more a function of international economic factors than of colonial institutions.

Acknowledgement

A slightly longer version of this paper was presented for delivery at the Anglo-Australian Science Meeting, the Royal Institution, London, on 7 January 1988. I would like to thank all participants for their comments and Stephen Nicholas for his subsequent thoughts on the script.

References

1. Thomas P. Hughes, 'Evolution of large technological systems' in W.E. Bijker, T.P. Hughes and Trevor Pinch (eds.), *The Social Construction of Technological Systems*, Cambridge, Mass., 1987.

2. For a general theoretical discussion see Paul Stoneman, *The Economic Analysis of Technological Change*, Oxford, 1983.

3. For a discussion of the British case see M. Kirkby, *The Decline of British Economic Power since 1870*, London, 1978; F.L. Payne, *British Entrepreneurship in the Nineteenth Century*, London, 1974; S.J. Nicholas, 'Technical education and the decline of Britain 1870–1914', in Ian Inkster (ed.), *The Steam Intellect Societies*, Nottingham, 1985. For a general theoretical introduction see J.W. Kendrick, *Understanding Productivity*, Baltimore, 1977.

4. For this case see the forthcoming Ian Inkster, 'Science, technology and the rise of the factory: an approach to institutional change in eighteenth-century Britain and Europe', *Journal of Economic Issues*, forthcoming.

5. See the general introductory survey in Chapter 5 of J.D. Gould, *Economic Growth in History*, London, 1972.

6. For a consistent emphasis on incrementation see Nathan Rosenberg, *Perspectives on Technology*, Cambridge, 1966.

7. For more details see Ian Inkster and Jan Todd, 'The support structure for the Australian scientific enterprise, circa 1851-1900', in Rod Home (ed.), *Australian Science in the Making*, CUP and Australian Academy of Science, Cambridge, 1988, pp. 102-133.

8. For such themes see essays in Nathan Reingold and Marc Rothenberg, *Scientific Colonialism, A Cross-Cultural Comparison*, Washington, 1987.

9. For Australian science and colonial imperatives see J.M. Powell, 'Conservation and Resource Management in Australia, 1788-1860', in J.M. Powell and M. Williams (eds.), *Australian Space: Australian Time*, Melbourne, 1975.

10. Home applies the phrase to physics: R. Home, 'The beginnings of an Australian physics community,' in Reingold and Rothenberg, op. cit., pp. 3-34 (p. 15).

11. The years being 1867-69, 1873-75, 1881, 1895-96; Victoria Patent Office, *Patents and Patentees* (Indexes), Vols. 1-28 (1854-93); *Official List of Victorian Patents and Trade Marks 1895-96*, series published annually, Melbourne. The latter is the same also for related data, and for the analysis of 1906.

12. Victorian Patents: 1076, 1214, 1307, 1210, 1308.

13. Ibid., 1185, 1179, 1240, 1252, 1814, 1831, 1872, 1893, 1986, 2008, 2024, 2026.

14. Ibid., 1734, 1870, 1902, 1978.

15. Ibid., 2944, 2948, 3005, 3012, 3023-4, 3054, 3079-88, 3137, 3140, 3015, 1991, 2982, 3006.

16. Ibid., 2991, 3022.

17. Ibid., all 1895-96.

18. Ibid., all 1895-96.

19. New South Wales Patent Office: *Indexes of Patent Applications in N.S.W., 1855-91*, published annually, Sydney.

20. Commonwealth Patents, nos. 5407, 5565, 6544, 6797, 6859, 7150-1.

21. Ibid., e.g., 5110-11, 5212, 8158-9.

22. Ibid., 5031-3, 5150, 5334, 5738, 5892, 6025, 7069, 7138.

23. Ibid., 5001, 5126, 5163, 5227, 5343, 5959-30, 6090, 6617, 7250-51.

24. Ibid., 5197.

25. Ibid., 5245, 5255, plus a series of earlier colonial lodgements.

26. Ibid., 5400-1.

27. Ibid., 5952, 6155.

28. Ibid., 5708.

29. Ibid., 6932, 6972, 5757-58, 7299-301, 7328-9, 7450-1, 7596, etc.

30. Ibid., 5460-2, 6964, 7332-4.

31. Ibid., e.g. 5006, 5369, 6299, 6934, 7273-40.

32. In progress under the title of 'Technical Training and Technical Change: The Inventors of Ballarat, 1850-1918'.

33. For the technological implications of this transition see: W. Little, *Guide to Ballarat*, Ballarat, 1890; H.J. Stacpoole, *Gold at Ballarat*, Kilmore, Victoria, 1971; Peter Milner, 'Gold mining and the development of engineering firms in Victoria', *Royal Hist. Soc. of Victoria, Journal*, Vol. 57, No. 4, Dec. 1986.

34. W.A. Bate, *Lucky City: The First Generation of Ballarat, 1851-1901*, Melbourne, 1978.

35. G.G. Turri, 'Australian mining and metallurgical patents', *Trans. Australian Inst. of Mining Engineers*, V, 1898.

36. Calculated from data in various issues of *Australian Official Journal of Patents*, Supplements.

37. G. Ranis, 'Industrial sector labour absorption', in *Economic Development and Cultural Change*, Vol. 17, 1973; J.C.H. Fei and G. Ranis, 'Less developed country innovation analysis and the technology gap', in G. Ranis (ed.), *The Gap Between the Rich and the Poor Countries*, New York, 1972; Ian Inkster, 'Prometheus bound: technology and industrialisation in Japan, China and India prior to 1914—a political economy approach', *Annals of Science*, Vol. 45, 1988.

38. *Historical Statistics of the United States, Colonial Times to 1970*, Vol. 2, Part 2, pp. 954-59.

39. To 1836 only aliens who had resided in the USA for two years could apply; from 1836 to 1861 aliens paid higher fees.

40. M.C. Uquhart and K.A.H. Buckley (eds.), *Historical Statistics of Canada*, Cambridge, 1965.

41. *Journal of the Society of Chemical Industry*, Vol. 23, 1903, pp. 341, 1157-58, *passim* (Canada sections).

42. W.R. Long, 'The chemical industry of Canada', read at Canadian Section, May 1903, in *JSCI*, ibid., pp. 527-37; J.A. Stovel, *Canada in the World Economy*, Cambridge, Mass., 1967.

43. New Zealand P.O., *Fifth Annual Report of Registrar of Patents, Designs and Trade Marks*, 1894.

44. *Ding. Poly. Journal.*, Vol. 295, 1894, pp. 160-64; *JSCI*, op. cit., Vol. 14, 1895, 406-9.

45. Ian Inkster, 'On "modelling Japan" for the Third World', *East Asia*, Vol. 1, 1983, Tables, pp. 168-170.

46. Ian Inkster, 'The ambivalent role of patents in technology development', *Bulletin of Science, Technology and Society*, Vol. 2, 1982.

47. D.M. Lamberton, *Science, Technology and the Australian Economy*, Sydney, 1970.

48. Tuvia Blumenthal, 'A note on the relationship between domestic research and development and imports of technology', *Economic Development and Cultural Change*, Vol. 27, 1979.

49. Angus Maddison, *Phases of Capitalist Development*, New York, 1982.

50. Lamberton, op.cit., p. 99.

Rational and Irrational Reconstruction

The London Sundial-Calendar and the Early History of Geared Mechanisms

M.T. WRIGHT

INTRODUCTION

The London Sundial-Calendar, a fragmentary instrument datable to c. AD 500 and combining a known type of portable sundial with a calendrical mechanism of toothed wheels, was acquired by the Science Museum, London, in 1983.[1] It has been described in a paper which dwelt upon the dating of it and outlined a conjectural reconstruction.[2]

The instrument joins a number of recorded examples of such sundials, and may well be the earliest known instrument to combine two (apparently) separate functions. But its great significance to the historian of technology is that it adds significantly to the exceedingly scarce evidence for the early history of geared mechanisms. It is the second oldest example of toothed gearing known (the oldest being the Antikythera Mechanism of the first century BC).[3,4] In an earlier paper it was also argued that, together with the closely comparable 'Box of the Moon', a calendrical device described by al-Bīrūnī in about AD 1000, it provides evidence for the transmission of technology from Byzantine to Islamic culture.

In the present paper I discuss further the reconstruction of the mechanism, both in its mathematical and its mechanical aspects; I also describe the methods adopted for practical reconstruction, relating them to evidence obtained from the original fragments. The insights gained in the course of both theoretical and practical reconstruction are shown to be of significance in interpreting the slender artefactual and documentary evidence for the early history of such mechanisms.

Most of what is to be said here concerns toothed wheelwork and geared mechanisms. The point has been made that in discussing the early history of toothed gearing one should consider the relationship between the high-technology tradition of mathematical gearing and the low-technology

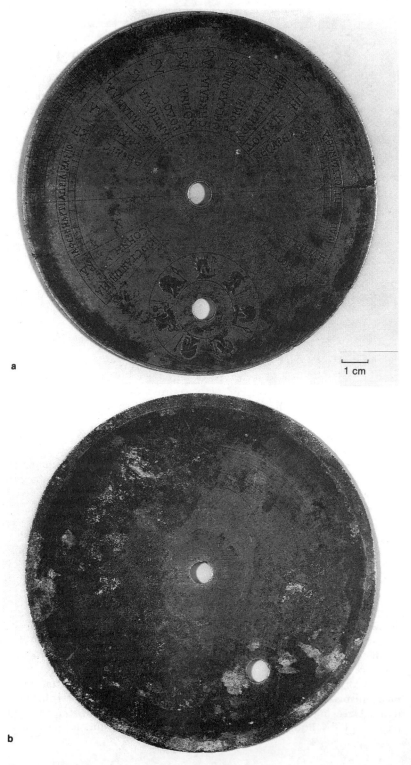

Figure 1 London sundial-calendar: dial plate. (a) Outer face; (b) inner face.

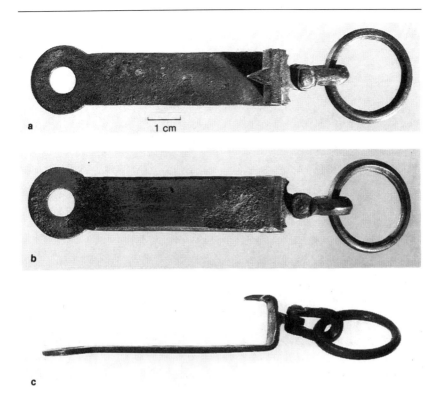

Figure 2 London sundial-calendar: suspension arm and ring. (a) Inner face and pointer; (b) outer face; (c) side view.

tradition of power transmission gearing; the former is characterized by the use of small wheels of metal, usually bearing triangular teeth, while the latter is characterized by the use of wooden wheels with wooden peg-teeth.[6] In as far as it is concerned with one or the other specifically, the present paper relates to the tradition of mathematical gearing and to the practice of making small gears out of sheet metal.

EVIDENCE FROM THE ORIGINAL FRAGMENTS

The original fragments are illustrated in Figures 1a–4b. Although they were described in the earlier paper,[7] it is appropriate to do so again here, laying greater emphasis on the evidence that they yield of the methods of manufacture employed. Figures 1a and 1b show a disc some 153 mm in diameter, which formed one face of the instrument (for convenience, the 'front'). On one face (Figure 1a) it bears a symmetrical double-sector declination scale with abbreviations of the names of the months of the Julian calendar, a latitude scale of degrees from 0 to 90 around one quadrant of the edge, and a table of place names with their supposed latitudes. These

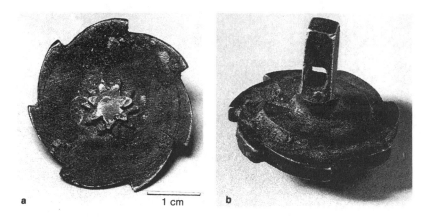

Figure 3 London sundial-calendar: input mobile. (a) Ratchet and pinions; (b) arbor.

Figure 4 London sundial-calendar: Moon disc mobile. (a) Display face; (b) rear view showing spacing washer and pinion.

features are connected with the instrument's function as a sundial. It also bears around the offset hole a ring of seven heads representing the rulers of the days of the week, which is connected with the calendrical mechanism.

Figures 2a–c show the arm and ring by which the instrument is suspended when used as a sundial. Figures 3a and b show a mobile comprising a washer, a ratchet wheel, and two pinions (of seven and ten teeth) riveted onto an arbor, which was intended to work in the offset hole of the large disc (Figure 1).

Figure 4a shows a 'Moon disc', a wheel of fifty-nine teeth designed to provide a display of the approximate shape and age of the Moon. It bears the Greek numerals 1 to 29 followed by 1 to 30 around it, to correspond to two months of 29 and 30 days; the face was brightened by tinning. The appearance of the display is seen in the reconstruction (Figure 11). Behind the disc (Figure 4b) are a spacing washer and a pinion of nineteen teeth; these parts were riveted onto a central arbor with a projecting pivot. The rivet is seen in Figure 4a.

The variations in composition as determined by X-ray fluorescence (Table 1) do not preclude a common origin, even from a single melting in which the alloy was poorly mixed. Moreover, the technique reveals the composition within only a confined area and close to the surface, such that surface contamination or leaching can affect the result. The presence of lead and tin from tinning or soft soldering, and the reduction of zinc content by corrosion are highly probable. The material may be described as 'red brass'. (Our more common 'yellow brass' contains more zinc.) The additions of tin and lead, if present throughout the mass, have contrary effects, tin hardening the alloy and making it more yellow, lead softening it and making it more red, while also making it flow better in casting. In the proportions found the effects would be small but those due to the lead would predominate.[8]

Such an alloy is soft when first prepared, and for mechanical purposes it is customary to harden it by hammering or other 'cold working'. The visible presence of a gross dendritic structure on the large disc and on the largest toothed wheel (Figures 5a,b) indicates, however, that these parts at least have not been subjected to cold working since cooling slowly from either their original casting or from an annealing or soldering operation at high temperature. Therefore the parts are probably soft and it is not surprising that they show considerable evidence of wear, both from use and from handling.

The majority of the flat parts—the main disc (Figure 1), the main part of the arm (Figure 2), the Moon disc and the washer behind it (Figure 4), and the pinions of seven and ten teeth (Figure 4)—are all of roughly one thickness, between about 2.2 and 2.3 mm. This suggests that they may have been cut out of one sheet that had been cast or worked to this thickness. The pinion of nineteen teeth (Figure 4b) and the piece on the arm to which the shackle is riveted are only slightly thicker at 2.7 and 3.0 mm respectively. It is possible that the components were cast individually as flat blanks, but there is no evidence that any surface has been left in an 'as cast' state. Such

Table 1 Chemical composition of the fragments, found by X-ray fluorescence (XRF). Traces of iron were also detected in all components. (Results obtained by the British Museum Research Laboratory.)

	Percentage composition			
	Copper	Zinc	Tin	Lead
Sundial (front) disc	86	13	0.2	1.1
Arm	83	16	0.2	0.5
Arbor with ratchet	84	14	0.4	1.7
Moon disc	79	18	0.3	2.8

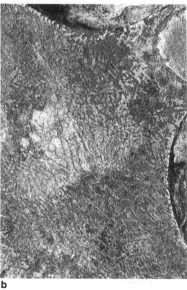

Figure 5 London sundial-calendar: dendritic structure of fragments. (a) Dial plate (enlargement from Figure 1a); (b) Moon disc (enlargement from Figure 4a).

evidence if found would be interesting as suggesting the manufacture of such instruments on a considerable scale.[9] Even so it is very doubtful that details such as holes or gear teeth would have been cast in.

The roughened zone seen around the edge of the large disc (Figure 1b) seems to testify to the former presence of a rim, presumably attached by soldering. (The degree of patination of the surface precludes the testing of this part of the surface by X-ray fluorescence.) My reconstruction requires the presence of such a rim fixed to the disc. Thus this feature provides further evidence that the components were cut from a sheet; had the disc been cast individually it would have been more convenient to have cast the rim integrally with it.

The circular marks seen on the rear of the disc within the roughened zone may be interpreted as evidence that the disc was turned in a lathe (probably between 'dead centres' on a mandrel passed through the central hole). On the other hand such marks could originate from the scraping or scouring of the plate by running a tool or block of abrasive stone around inside the rim. Moreover, the periphery of the disc is markedly out of round. The only piece which seems to show clear evidence of having been finished in a lathe is the suspension ring (Figure 2). All other features of the original fragments are compatible with the use of a modest repertoire of simple techniques using hand tools.

The face of the large disc (Figure 1a) was probably laid out using simple geometrical methods. Four equally spaced marks made with a prick-punch or awl on the outermost circle form the basis of the layout. The small circles surrounding the two holes could have served the workman as a guide in opening the holes to size, to ensure that the holes were not displaced. The modern (but traditional) workman makes such circles, emphasizing them if

Figure 6 London sundial-calendar: markings on dial plate (enlargements of Figure 1a). (a) Graduations of declination scale; (b) graduations of latitude scale; (c) discontinuities of arcs.

need be with punch-marks. The outermost circle may similarly have served as a guide in filing the edge.

The radial lines marking the centre line and boundaries of the declination scale, and the divisions between months of this scale and of every 5° of the latitude scale, were put in with some nicety using a straight edge. The lines between the heads for the days of the week and those between the place names were put in equally neatly but less care was taken to make these divisions equal. The subdivisions of the months into three parts and of the 5° intervals into single degrees have, however, been put in more lightly freehand (Figures 6a,b). Perhaps these represent a later 'improvement'.

There is nothing that shows clearly whether any of these lines was put in using a template or division plate. But there is an odd feature of the arcs bounding the declination scales and the table of latitudes. At opposite ends of a diameter there are discontinuities: corresponding arcs on either side of the diameter are struck with the same radius but with their centres displaced by some 0.4 to 0.6mm along the diameter (Figure 6c). If the arcs were struck with compasses, it is odd that, even supposing a double or an elongated centre mark on the disc, the same centre was taken for all the arcs in one semicircle while the other centre was taken for all the arcs in the other semicircle. A more plausible, though surprising, explanation is that a rather complicated template could have been used.

Figure 7 shows a conjectural reconstruction of such a template. The error observed on the dial plate could arise from a defect in the template or from its being positioned without due care. I advance this suggestion with considerable diffidence because the use of such a template would seem to save very little effort in actually laying out a dial (while considerable effort would be involved in making it), to promise no improvement over the use of the compasses, and to indicate a commitment to the making of a considerable number of such dials, all of this size.

All the lines, arcs and radii, are scribed. The lettering is all engraved. It is amusing to note that the engraver mistakenly put in the latitude ΛB (32°) for Dalmatia, later correcting it to MB (42°).

The heads, if engraved, are remarkably deeply cut. Patination within the depressions makes investigation difficult but it seems likely that these designs were firstly formed roughly by punching with simple part-punches and then detailed by engraving. This combination of techniques was already commonplace in the sinking of dies for coining.[10] While the lettering is fairly neat, the execution of the heads is exquisite and it seems possible that different workmen performed these two tasks.

The arm (Figure 2) is bent up from the flat and the piece on which the U-shaped shackle swivels has tangs riveted into holes in it. As described below, its oblique setting is probably the result of accidental damage. The profile of this part and the stopped chamfers on the straight part of the arm show that the maker had a fine eye for pleasing detail.

The smaller pinion, of seven teeth (Figure 3a), has been bruised out of shape, perhaps in the act of forming a rivet on the end of the arbor at its centre. The undamaged teeth, and those of the larger pinion of ten teeth, are of a severely triangular profile and an arc of a scribed circle is visible at their

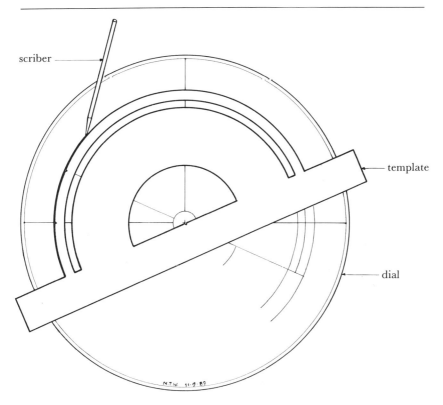

Figure 7 Template for laying out dial plate: conjectural reconstruction.

roots on the smaller one. A similar scribed circle and crude radial divisions are visible on the ratchet wheel, strongly suggesting that it was filed, not cast, to shape. The washer on the other side of the ratchet wheel has a piece broken out, perhaps at least in part the consequence of it having been forced onto the arbor when too tight a fit. This part is thinner than all the other flat parts and may have been thinned by hammering, which, if carried too far without annealing, would have made it brittle. The ratchet itself is significantly thicker than all the other flat parts; there is no obvious reason for this.

The wheels of fifty-nine and nineteen teeth (Figures 4a,4b) also have triangular teeth and circles scribed at the roots. Both also show radial divisions running to the tips of the teeth, evidence of the division of the wheels prior to cutting the teeth. On the Moon wheel the radial divisions also serve, with another scribed circle, to define 'boxes' within which the numerals are engraved, one to each tooth space. Some of these radial lines are doubled, showing that the workman blundered in his work. Near the centre of the Moon wheel, between two roughly circular openings, are traces of two scribed circles showing that the openings were laid out by scribing four equal touching circles. The scribed layout lines indicate that neither the

Figure 8 London sundial-calendar: the two mobiles positioned on the dial plate.

Figure 9 London sundial-calendar: all four parts assembled (wooden support modern).

teeth nor the openings were cast-in features. The formation of the teeth is considered in detail below. The lack of roundness of the openings is taken as evidence that they were filed to size, not drilled or bored in a lathe.

The face of the disc was brightened by tinning, traces of which remain in the scribed lines (which must therefore have been made first).

The face of the nineteen-toothed pinion (Figure 4b) exhibits a circular sink. This feature might have been formed by casting or by turning in a lathe or by drilling; the presence of corrosion products makes it impossible to decide the point. Nor is its purpose clear; there is a shoulder on the arbor which is thereby sunk below the flat face, but if the pivot on the arbor were made larger there would be no need of a distinct shoulder. The shape of the rivet on the other end of the arbor (Figure 4a) suggests strongly that the shank of the arbor is square, presumably threaded through square holes to prevent the wheels turning on it. Square arbors and square holes are found in the Antikythera mechanism.[11]

Figures 8 and 9 indicate how the four fragments would be juxtaposed according to my reconstruction. The pinion of seven teeth engages the Moon wheel of fifty-nine teeth. The arbor with the pinion is supposed to be moved on one-seventh turn for each day, moving the Moon wheel on by one tooth, that is, one day. The arm, the pointer of which (Figure 2a) lines up with the latitude scale (Figure 1a), allows room for a 'back' dial plate parallel to the surviving one (Figure 9). Figures 8 and 9 may be compared with the views of my reconstruction (Figures 10 and 11).

THE FORM OF THE INSTRUMENT: MINIMAL RECONSTRUCTION

By a 'minimal reconstruction' I mean one that accounts for all the available evidence in the simplest possible way. The design and execution of the reconstructed parts must be as plain and as simple as possible. Ideally it amounts to an *absence* of design; the most important tool in making a minimal reconstruction is Occam's razor. Such a reconstruction may be compared to the piecing together of fragmentary pottery, filling the lacunae with material of a neutral tint. Where it is necessary to introduce features, patterns and precedents must be sought in the surviving fragments or, failing that, in comparable historical material. To extend the analogy of the pot, any decoration of the restored portions should extend the scheme of the original parts or copy the design of a comparable specimen. My reconstruction (Figures 10 and 11) is certainly close to minimal in its exterior features. The size and shape of the body are dictated by the 'front' disc (Figure 1) and the arm (Figure 2). There must have been a 'rear' disc, to have the necessary two holes for the Moon disc (Figure 4a) to be viewed through. A circular rim forms the simplest way of separating the two and forming a single body. It would have to register positively against the front disc or be fixed to it, and the evidence of the roughened zone on the back of the disc (Figure 1b) and the absence of other features suggest it was fixed. The rear disc must then have been removable but must have registered positively like the 'climates' of an astrolabe, since part of the internal mechanism (the Moon disc;

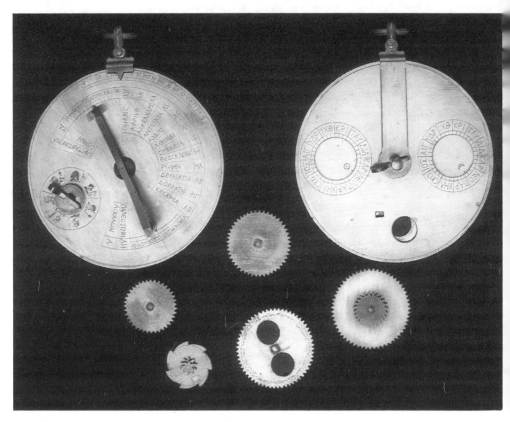

Figure 10 Reconstructions by M.T. Wright: two complete instruments and a set of wheels made entirely by hand.

Figure 11 Reconstruction by M.T. Wright. (a) Sundial face; (b) calendar face.

Figure 4) must have been attached to it while another part (the other arbor; Figure 3) was attached to the front. The design of the other displays seen on the rear will be discussed below but it will be appreciated here that the style in which they are executed is copied from the plainer parts of the front. The vane which combines the functions of gnomon and scale of the sundial is copied from other recorded examples. But in fashioning the two cotters, one in the stem of the vane and one in the projecting end of the 'day-of-the-week' arbor (Figure 3) I abandoned the severity of 'minimalism' for shapes derived from the Arabic name for such components on astrolabes, *faras* (literally 'horse').

A 'MINIMAL RECONSTRUCTION' OF THE LOST WHEELWORK

As described above, the surviving fragments establish that the geared mechanism was calendrical, the pinion of seven teeth engaging the wheel of fifty-nine teeth. Functions have to be found for the other toothed wheels, the pinion of ten teeth turning in seven days (Figure 3a) and the pinion of nineteen teeth turning in fifty-nine days (Figure 4b). Accordingly, further calendrical functions are sought which could be driven by trains of gears leading from these two wheels.

The four surviving toothed wheels, combined like this, correspond exactly to parts of an instrument, 'The Box of the Moon', described by the Islamic polymath al-Bīrūnī in about AD 1000.[12] In this, a pointer moving over a dial showing the days of the week drives a Moon wheel turning once in two lunations, precisely as described above. (In al-Bīrūnī's mechanism there is no ratchet.) It also drives (directly) a mobile indicating the position of the Moon in the zodiac and (indirectly, through the Moon wheel) another indicating the position of the Sun in the zodiac. These latter two displays can equally be regarded as indications of the sidereal month and the year respectively. The full gearing scheme may be written as 59/7 × 59/19 × 48/24 for the year and 40/10 for the sidereal month. (In this notation, each fraction represents a pair of meshing wheels. The driver is the denominator of the left-hand fraction. Thus 7 is turned by hand and drives the Moon disc, the 59 above it. Nineteen, the denominator of the next fraction, is the pinion on the back of the Moon disc, and so on. In the second expression 10 is the driver, which is of course rotated with the 7 as in the London sundial calendar, Figure 3b). In each case, working out the arithmetic and multiplying by the period of the first wheel of the train, seven days in each case, gives the period of the last wheel—366.42 . . . and 28 days respectively.

The suggestion has been made,[13] with rather little justification, that the missing wheels of the London sundial-calendar might be reconstructed according to an 'improvement' of al-Bīrūnī's scheme; the argument for doing this is given here at greater length.

The correspondence of the 7 and 59 is not surprising, arising from the calendrical data adopted, that is the seven day week and the good approximation of 29½ days to one lunation, and from the basis of allowing the displacement of one tooth-space to one day. So this correspondence alone is not enough to justify reconstruction of the missing mobiles according to al-Bīrūnī although it is remarkable that the other wheels present, the pinions of 10 and 19 teeth, do also correspond.

The surviving fragments establish that the mechanism is calendrical and one must look for likely calendrical functions to be worked from the two unused wheels. In both cases the search is confined to step-down ratios, both on account of general mechanical principles and in view of the particular constraints imposed by the surviving parts. It is sufficiently obvious in the case of the pinion of 10 that the driven wheel cannot be smaller; and for the pinion of 19 I find in practice that it would be impossible to fit a small wheel engaging with the pinion of 19 and carrying on its arbor a further wheel or pinion and a pivot, without it fouling some part of the existing mobile. In this way it may be deduced that the smallest wheel that could engage the pinion of 19 would probably be an equal wheel of 19. More generally, wheelwork with such crude teeth works much better in step-down than step-up ratios, especially when the driving member has a low number of teeth.

The existence of the pinion of 19 prompts consideration of the Metonic ratio, which was well known when the instrument was built. For the present purpose it is expressed as 19 years = 235 lunations. The pinion of 19 turns in two lunations, so if it engaged a wheel of 117 or 118 teeth the latter would rotate in one year (approximately). The display would probably comprise an indication of the Sun's position in the zodiac (since lunar information is already displayed) or an indication of the Julian calendar, or both. Certainly such an indication seems a very likely one to have been included in such a calendrical/astronomical display, and the inclusion of a calendar dial that could be of direct use in setting the sundial does lend a degree of unity to these two separate parts of the instrument. It would be quite practicable to make a wheel of 117 or 118 teeth but it could not fit within the present instrument as envisaged; the biggest wheel which could be accommodated between the central pin and the side of the case would have about sixty-six teeth, and the biggest wheel which could ride on the central pin without fouling the existing mobiles would likewise have about sixty-six teeth.

Returning to the Metonic ratio, 19 years = 235 lunations; note that 235 = 5 × 47. Thus the missing train might be reconstructed as follows: 60/19 × 47/24, giving the Metonic ratio *exactly* between the pinion of 19 turning in two lunations and the wheel of 47 turning in one year. However, since two lunations is taken as 59 days, this yields a year of 365.0 days. The numbers 60 and 24 are arbitrary, and any pair of numbers in the ratio 5:2 would serve; using wheels all of one pitch (see below) this train can be fitted conveniently into the reconstruction.

Equally, two other trains noted earlier,[14] 60/19 × 49/25 and 60/19 × 51/26, can be accommodated to give less good ratios between lunation and year but improved values of the year of about 365.18 days and about 365.47 days respectively. Doubtless other trains are possible; I have not made an exhaustive search.

Any of the above can be made to fit with the arbor intermediate between the lunation display and the year display falling over the punch-mark, as already described. A train such as 60/19 × 47/24, which can be justified as embodying the Metonic ratio, can claim precedence over others which are not seen to have any such basis, but from the practical point of view there is little to choose between them. (For the division and cutting of gear wheels of any number of teeth, see below.) However, the closely similar train given by al-Bīrūnī, though embodying a less exact ratio, does offer some real practical advantages. The corresponding part of this train is 59/19 × 48/24.

The trains discussed above are fragmentary, representing the existing pinion of 19 teeth and the possible wheels leading from it. In each case the complete train includes the two-lunation Moon disc of 59 teeth. So we see that in al-Bīrūnī's train the two largest wheels are identical; in making them, whether by original division on the blank or by marking off from a division plate (see below), only one need be marked out and the two may be held together and cut as one. If a division plate is used, a circle of divisions is saved.

Similarly the two wheels of 24 and 48 teeth may be marked out from a single circle of 48 divisions on a division plate. Or, if they are to be divided originally, it is somewhat simpler to set out these numbers than, say, 25, 26, 47 or 49 or 51.

Thus, al-Bīrūnī's train may perhaps be regarded as a train especially simplified for the workman. The approximation in the relation between the lunation and the year, while poor compared to other trains given above, is still good enough so that the year (or Sun in the zodiac) indicator would not need setting more often than the lunation disc. Moreover, as will be argued below, there are indications that al-Bīrūnī, writing in about AD 1000, was describing a traditional device, reducing the chronological gap between his 'Box of the Moon' and our sundial-calendar. On these grounds, the train given by al-Bīrūnī was adopted in reconstructing this part of the London sundial-calendar.

Turning to the reconstruction of the other missing part of the wheelwork, that leading off from the existing pinion of 10, I again follow al-Bīrūnī as to the function of this part of the train. So far we have a display driven from the day-of-the-week dial to show the age and phase of the Moon, and the position of the Sun in the zodiac, and perhaps the date in the Julian calendar. Adding a display showing the position of the Moon in the zodiac represents a logical extension of this scheme. While it remains unclear what, if any, use the instrument as reconstructed might have been put to, elaborations of this 'minimal' reconstruction do yield some interesting possibilities which will be discussed below. It is not easy to imagine what other motion or calendrical function could fit with what we already have. When it comes to the numbers of teeth, however, the case seems different. Al-Bīrūnī has a wheel of 40 engaging the pinion of 10, so that his Moon in the zodiac display rotates in 28 days as against the true period of about 27.32 days. This is a poor approximation, requiring to be reset much more frequently than the other parts of the display. Moreover, it does not make sense: the same effect could be achieved by a wheel of 28 teeth engaging the pinion of 7 (and clearing the wheel of 59 which engages the pinion). Or, if the objective were to separate the arbors so as to allow larger displays, a pinion of 12 could have engaged a wheel of 48 to give the same result, conferring the advantages of making the wheel identical with another in the mechanism (see below) and making the pinion rather easily divided. On the other hand, if the pinion of 10 turning in seven days engages a wheel of 39 teeth, the resulting period is 27.3 days (exactly), a good approximation to the lunar sidereal period of 27.32 days, comparable in accuracy with that adopted for the lunation. Although 40 is a round number, it is not a particularly convenient one for original division, requiring that the eighth part of the circle should be divided into five parts by trial, as described below.

In reconstructing the London sundial-calendar I have, then, adopted al-Bīrūnī's train for the Sun-in-the-zodiac display and a modification of it for the Moon-in-the-zodiac display. The trains leading each way from the input arbor are as follows:

59/7 × 59/19 × 48/24 (output turns in 366.42 days)
39/10 (output turns in 27.3 days)

Figure 12 shows how this wheelwork fits within the reconstructed instrument.

So far as making a practical working instrument is concerned, minor variations as discussed above make no difference. One could adopt al-Bīrūnī's numbers throughout, or one could deviate from them, for example to incorporate the Metonic ratio; either way the wheelwork would fit and the instrument would work. To show that the instrument could easily be made by simple methods—one of the major objectives of practical reconstruction—it did not matter which scheme was adopted.

The uncertainties over the particular numbers of teeth notwithstanding, the reconstruction described is the simplest I have found to account for the gears of the surviving fragments. I therefore regard it, at least provisionally, as 'minimal'.

MECHANICAL ARRANGEMENT AND DETAILS

The external form and general arrangement of the instrument, and the general scheme of the wheelwork, have already been discussed.[16] It remains to consider how the wheelwork and other internal details are arranged. Most of these details are necessarily conjectural. The paucity of comparison material has been noted, but ancient parallels have been followed where possible. Yet as with the exact numbers of teeth on the restored wheels, many of the details could be varied without invalidating the reconstruction as a whole. They were all sufficiently 'obvious' that most of them were thought out before practical reconstruction began and the remainder fell in naturally during construction without any need for preliminary drawing or trial.

The 'input' arbor must have been held up to its pivot hole in the 'front' dial plate. Under the scanning electron microscope a circular pattern of wear is seen on the dial around this hole. The most obvious explanation is that there should be a washer of this size and of a thickness to hold the mobile without undue shake when fastened by a cotter through the cross-hole in the arbor. The cotter also doubles as a pointer on the ring of heads (Figure 11a).

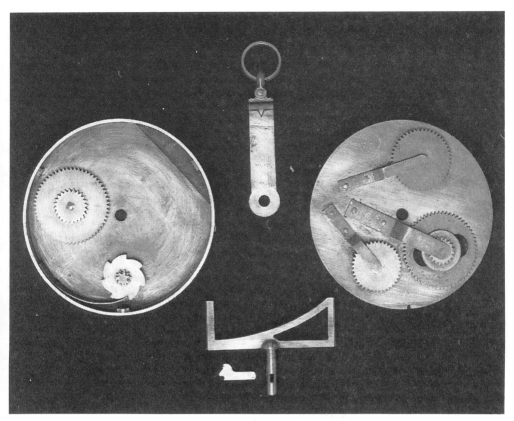

Figure 12 Reconstruction by M.T. Wright, internal details. Left: box formed on rear of sundial face; right: plate carrying calendrical display.

Figure 13 Reconstruction by M.T. Wright: detail of pivot for intermediate wheel.

A click-spring to work in the ratchet is provided by fastening a suitably shaped strip of springy metal to the side of the case (Figure 12).

The arrangement of the Moon disc behind two openings in the rear dial follows from the geometry of the disc itself. It corresponds to al-Bīrūnī's description,[17] and to the arrangement found in the comparable calendrical device built into the back of an astrolabe by Muḥammad b. Abī Bakr ar-Rāshidī al Isfahānī, dated AH 618 (= AD 1221/2), now in the Museum of the History of Science, Oxford.[18] The pivot on the back of this mobile is carried in a bent up finger or 'cock' riveted to the dial plate (Figure 12). Since the piece has only the one pivot it is really necessary that it is held up to the plate in some such way. For the displays representing the motion of Sun and Moon in the zodiac, one could follow al-Bīrūnī in having pointers, or the Oxford instrument and perhaps the Antikythera mechanism in having rotating rings or discs flush with the dial.[19] Pointers would be liable to be fouled by the suspension arm and flush rotating discs were adopted. Each mobile comprises a disc to fit the hole in the dial, a larger disc behind it, and the gear wheel, riveted together by a central pin. The first and second discs form a shoulder on which the mobile takes its bearing, needing only to be held up to the dial, which is done by a bent-up finger riveted to the plate.[20]

The remaining intermediate mobile is made with a pivot towards the front dial plate which is run in a collar soldered to that plate in the position indicated on the original by a large pit, apparently punched or drilled; compare Figure 1b and Figure 13.

It was this singular but apparently purposeful feature which was taken as a hint for running this mobile in a bearing fitted to this plate. It is conceivable that the mobile had two pointed pivots, one running in this pit and the other in a corresponding one in the rear dial, but no parallel for such an arrangement is known from before the late medieval period. Moreover it would be troublesome to assemble the instrument and undoubtedly such pivots would give trouble. The arrangement of a pivot running in a collar soldered to the plate is taken from al-Bīrūnī's description.[21] However this mobile was pivoted, an arrangement (such as either of those suggested above) in which it is not attached to the rear dial has an interesting consequence: when the rear dial is lifted out the three mobiles providing the display (which as reconstructed are attached to it) may be turned independently for resetting. It would be necessary to do this at fairly frequent intervals and a reconstruction of the instrument which would allow the owner to do this simply without the need for appreciable manipulative skill is particularly convincing in this context.

Some form of counterpoise to the wheelwork is required if the instrument is to hang upright when used as a sundial. There is ample room in the case (Figure 12).

The only inscriptions engraved on the rear dial (Figure 11b) are abbreviations of the names of the zodiacal constellations in Greek and simple symbols for the Sun and Moon on the rotating disc. Almost certainly other lettering would have been present on the original.

THE MANUFACTURE OF TOOTHED WHEELS

General Considerations
It has been supposed that making toothed wheels for such mechanisms is difficult or demands the use of special apparatus. Whether or how far specially developed techniques are required depends on the demands made of the wheelwork, and the scale on which it is constructed. None of the evidence now available for ancient geared mechanisms need lead us to question the accepted view, that wheel-cutting apparatus was introduced in the late sixteenth or seventeenth century; that its introduction was gradual; and that at first the wheels were only slit in the wheel engine, the teeth being 'rounded-up' (shaped) by hand with a file.

In making a gear wheel, the workman has to solve three basic problems: firstly the size of the wheel must be chosen; secondly the edge must be divided into the correct number of intervals; and thirdly the teeth must be formed according to these divisions. Sizing wheels can be more subtle than might at first appear, and this point will be taken up later. It is connected with the further problem of *pitching* the pairs of the wheels at a suitable separation from one another.

The fundamental idea is that wheels (without teeth) placed in rolling contact will rotate at rates inversely proportional to their diameters or radii. Rolling wheels are mentioned in passing in an Aristotelian text of the third century BC,[22] which treats various mechanical matters in terms of levers; there is, however, no discussion of the velocity ratios of such pairs of wheels, unless that is what lies behind the obscure discussion of pairs of wheels fastened together.

Once the idea of unequal wheels rolling together had been developed, teeth could have been regarded as a deliberate roughening to prevent slippage. Otherwise, it is possible that the starting point was the functioning of primitive wheel teeth themselves, seen as levers, leading to the proportioning of the radii of the wheels. On a more practical level, to function well the teeth must be (roughly) equally spaced on the two mobiles; thus a wheel to contain (say) twice as many teeth must have a circumference, and so a diameter, twice as great. It is perhaps a question of whether the origin of toothed wheels is to be regarded as a creation of the mathematician or of the mechanic. In either case, the mechanic (nowadays) tends to think in terms of the numbers of teeth in the wheels, while recognizing that the sizes of the wheels (defined in some suitable way) are proportional to these numbers.

Division of Wheels
It is the division of the wheel—the setting-out of the tooth spaces around its circumference—which has commonly been held to be the most difficult operation. The problem of dividing a circular dial is identical and my remarks apply equally to this task.

There are mechanical methods of dividing the work into any number of parts using very simple means. A commonly cited way is in effect that described by John Smeaton as Henry Hindley's method:[23] equal intervals are stepped off along a flexible band which is then joined to form a loop of the desired number of intervals; a circular disc is reduced until the loop of band may be fitted around it, whereupon it becomes a division plate, pattern or protractor. This and other methods could have been available to the ancients, but it remains to be proved that they were even thought of. The use of some sort of guide might, for example, be demonstrated by the repetition of certain errors of division from one dial or wheel to another. The use of a pattern or guide would certainly be convenient to a workman who had a number of similar pieces to make, and a relatively large division plate could be divided, even by trial, with more nicety than a smaller dial or wheel, so that thereafter dials or wheels could be laid out both better and faster. However, it is not necessary to suppose that any such equipment was available. What follows is intended to show that the task may be performed quite adequately and not too tediously with the simplest means.

Certain divisions of the circle may be obtained by geometrical construction: the circle may be divided into four by constructing perpendicular diameters, or into six by stepping the radius round, and the arcs may be reduced by repeated bisection. Division into thirty-two or forty-eight, for example, may in principle be carried out with geometrical exactitude. All other numbers involve the division of arcs by approximate means. Thus in dividing the circle into sixty parts, one can divide it into six and thence into twelve, but these arcs must be quinquesected. One may approximate by quinquesecting the chord and drawing radii through the divisions, or one may by trial find the opening of the compasses which will step five times into the arc, for example.

Other numbers of divisions may best be obtained by more oblique means. The number thirty-eight is found in the Antikythera mechanism; it may be approached as follows:

The circle is divided into six and one sextant is subdivided by bisection into twelve. It is easy to trisect one of these arcs by trial, so that the circle could be divided into thirty-six; but we require smaller divisions so that there are nineteen, not eighteen in each semicircle, so the compasses are closed a little further and stepped round seven times. The compasses should now straddle the division at the end of the sextant by one-third of their opening, which is easily judged by eye. If they do not, the compasses are adjusted and another trial is made, then if the result of this is satisfactory the compasses are stepped round more firmly, checking at the end of each sextant. It is found that the size of the step can vary a little with how firmly the point is pressed into the metal, but with experience this can be allowed for. After three sextants (a semicircle) the compasses should of course fall in with the original division into six. If they do not fall in closely enough, the compasses may be stepped back the other way from that mark and the later marks from the first set, with the greater accumulated error, ignored. It is wise, in any case, to put these marks in only lightly and to deepen them subsequently with a prick-punch or scriber and straight-edge. The second semicircle is treated in the same way. Similar procedures may be developed for all numbers.

Price has laid great emphasis on the distinction between 'easy' and 'hard' numbers, to the extent that he has preferred 'round' numbers to others and even numbers to odd in his reconstruction of the Antikythera mechanism.[24] He also supposed that al-Bīrūnī's scheme, involving the numbers seven, nineteen and fifty-nine was only an 'armchair invention' for the same reason. (It must be said that he was quite ready to retract this suggestion when shown the fragments of the London sundial-calendar shortly before his untimely death.) It will be appreciated that a round number such as sixty is but little more convenient than, say, fifty-nine; both involve trial stepping or similar non-geometrical procedures. Also, although numbers such as forty-eight and sixty-four may be produced wholly by strictly geometrical procedures, the results of proceeding in such a way, with their accumulated errors, are no better on the scale of these geared mechanisms, and are achieved more slowly, than the results of geometrical division into manageable arcs followed by subdivision by trial stepping.

The foregoing account indicates that the supposed difficulty in laying out 'odd' numbers of divisions is largely illusory. This has been confirmed practically by making a set of wheels for the reconstruction of the London sundial-calendar by hand, as described below.

It was mentioned above that the use of some form of dividing plate would be a convenience to the workman who had many wheels or dials to lay out, by speeding his work up; it could also confer the advantage of increased accuracy when, as is usual, the division plate is larger than the piece to be divided. Against this, the transfer of the divisions can be associated with characteristic errors, such as those due to lack of concentricity of the workpiece and the division plate, and any errors present in the original division of the plate will be reproduced in all workpieces made from it. If the workman had a division plate it would obviously suit him to use the numbers on the plate and their factors in preference to others wherever possible. It has been noted that al-Bīrūnī's train for the 'Box of the Moon' (using two wheels of 59, a 48 and a 24, and perhaps a 40 and a 10) may be regarded as rather convenient in this respect. In analysing the probable number of teeth on the wheels of the Antikythera mechanism, the argument that the same numbers or their factors are likely to recur is probably stronger than Price's argument for 'round' numbers.

There are ways of using a division plate or other divided circle to produce numbers of divisions slightly different from those laid out on it, to a reasonable approximation. Suppose, as an example, that one wished to lay out a wheel of thirty-eight teeth but that the nearest number on the plate was forty. One could scribe a diameter on the wheel and then set the wheel off-centre on the division plate so that the ends of the diameter corresponded to (say) divisions numbered 1 and 20 instead of 1 and 21 as they might if the wheel was centrally placed. Lines can then be drawn radially to either the wheel or the plate through divisions 2 to 19 of the division plate to divide the semicircle into nineteen parts. The wheel would then be readjusted to divide the other semicircle. Similar techniques could be used for dividing, say, one quadrant or one sextant of the wheel at a time. In each case the divisions would show a characteristic pattern of errors from which in principle the method used to divide it could be deduced. In practice, the errors due to this cause would probably be small compared to others present.

This is obvious when the circle of divisions on the plate contains nearly the number of divisions wanted on the wheel, as above. Bromley and Percival have suggested a powerful extension of this method using a small divided circle to place divisions in one sector of the wheel at a time.[25] The circle is positioned to give the correct number of intervals in the sector and lines are drawn radially on the wheel through the divisions on the small circle. If the circle is small enough to pass through the centre of the wheel to be divided, then the division of the sector is *exact* because the equal arcs on the divided circle subtend equal angles at any point on its circumference. It follows that if the divided circle is of any size from about that of the wheel down to about that which fulfils this condition the geometrical errors of this method of division are rather small. These are, of course, compounded with the errors of division of the workpiece into sectors, errors of division of the divided circle, errors of placement and errors of transfer of the divisions in ruling the radial lines.

If it could be shown that division plates or patterns had been used in making early components, this would reveal something of the outlook of the workman and perhaps yield information on the likely scale of operation. It has been suggested that the bronze disc illustrated in the catalogue of the Parisian instrument dealer, Alain Brieux,[26] and now in The Time Museum, Rockford, Illinois, is such a division plate but this identification remains doubtful.

The essential point, however, to emerge from the foregoing discussion and the practical trial to be described below is that, failing positive evidence, it is not necessary to suppose that the ancients used any form of division plate at all.

Formation of Wheel Teeth

In the Antikythera mechanism, the London sundial-calendar, and a number of later mechanisms, as well as in early manuscript sources, the wheels have teeth with straight flanks and more or less sharp tips. These teeth are usually described as 'triangular', not infrequently as equilateral triangles. Price so described the teeth of the Antikythera mechanism although in view of their small size and the means of examination open to him it is doubtful whether he can have intended this observation to be taken literally.[27] He supposed that this form was arbitrarily chosen by a designer who was a mathematician or geometer, but there is a practical point of view according to which it is more significant to consider the shape of the spaces between teeth. The obvious way of producing such teeth with simple modern tools is to form each space with a triangular file, such as those used for sharpening saws. Both saws with pointed teeth and files survive from the Roman period, so saw files of triangular section or with a similarly acute edge very probably existed. If they did not, some other means of forming a notch in a metal sheet, such as a slip of abrasive stone worked into a triangular or other section with similarly acute edges, must have been used. Without presuming a uniform state of technical development, whether geographically or through the period covered by the objects that we are considering, it seems safe to suppose that, in any centre of metalworking sufficiently advanced to have produced such a geared mechanism, suitable files or stone slips for cutting the teeth would have been known.

The natural tendency in forming teeth in this way is to sink the file directly into the edge of the disc, forming a notch which is the counterpart of the section of the file. Whatever the angle of the edge of the file, the spaces would be the same and the teeth themselves would be somewhat less, depending on the number of teeth in the wheel. This is what is observed in the wheels of nineteen and fifty-nine teeth of the London sundial-calendar; not only are the spaces of the two wheels of the same angle, but on both wheels they show the same asymmetry near the root, indicating an imperfection in the formation of the file. To operate on the two flanks at either side of a space separately, as one would have to with a knife file (that is, one formed somewhat like a knife, with an acute edge) or if one wished to bring the teeth of different-sized wheels to the same angle, would make the work much slower and would make it harder to keep the results even. With small numbers of teeth, however, as with the pinions of seven and ten in the London sundial-calendar, the angle of the space must be enlarged in order to strengthen the teeth and the centre of the pinion. Significantly, in the London sundial-calendar, the teeth of these pinions and the spaces between them are seen to be far less uniform in size and angle. It will be observed that, if the spaces are of uniform angle, wheels with smaller numbers of teeth have teeth that are not only narrower in angle but longer radially. This has implications for the sizing of the wheels.

Sizing and pitching toothed wheels

In sizing toothed wheels the modern mechanic employs the concept of the pitch circle representing the effective diameter of the wheel. A pair of wheels, correctly sized and with teeth of suitably matched profiles, behave much as though the two pitch circles roll on one another without slipping. The pitch circles are usually intermediate in size between the root and tip circles, defined by the bottoms and tops of the teeth, their precise diameters depending on the profile of the teeth and the *pitching* or separation of the wheels. The diameters of the pitch circles are proportional to the numbers of teeth.

In practice the *lead* is never as uniform as this ideal picture suggests; that is, the velocity ratio of the two wheels varies as the individual teeth run through their engagement. The simple geometry outlined above is also departed from quite deliberately for various reasons. For example, where a pinion leads a wheel, it is found advantageous to make the pinion somewhat oversized to obtain a smooth action.

In considering ancient wheelwork, we should not presume that the designer made use of any such concept as the pitch circle. The wheels may have been sized according to their tip or root diameters, and some allowance, by 'rule of thumb' or otherwise, may have been made to obtain a suitable action. It seems particularly unfortunate that in analysing the Antikythera mechanism, instead of giving tip or root diameters that could be measured (in so far as measurements may be got from the fragments at all!), Price gives diameters for the wheels which he calls 'pitch diameters' which

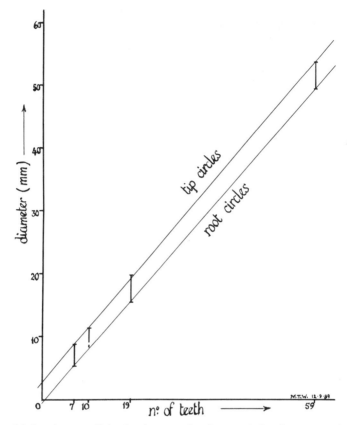

Figure 14 London sundial-calendar: graph of root circle diameters of wheels against numbers of teeth.

depend on estimation; he does not state where he takes the pitch circle to lie.[28] If the radiographs from which he worked were accessible, this work could with advantage be revised.

The fragments of the London sundial-calendar include four toothed wheels, of seven, ten, nineteen and fifty-nine teeth and all of approximately one pitch. On three of these, scribed root circles are clearly visible (the pinion of ten teeth is largely obscured by the pinion of seven).

Within the limits of the uncertainties of the measurements, it seems that the diameters of these root circles were proportioned to the numbers of teeth that the wheels were to contain (see graph, Figure 14). A consequence of this sizing technique is that the smaller wheels are oversized, although in the very small wheels (seven and ten teeth) the effect is moderated by increasing the angular size of gaps and teeth, thereby also strengthening the teeth. In a mechanism in which the pinion leads the wheel, as in the reconstruction of the London sundial-calendar, the results of such sizing are entirely satis-

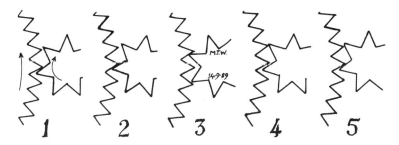

Figure 15 The engagement of triangular wheel teeth.

factory. The diameters of the root circles of all the wheels of the reconstruction were determined in this way. The action of these trains of wheels, crudely sized and with crudely formed teeth as described, was found in practice to be entirely satisfactory. The nature of the engagement of a pair of such wheels is illustrated in the series of diagrams of Figure 15, and the graph of Figure 16, in which the velocity of the driven wheel is shown assuming the driving pinion turns at a constant rate. These figures illustrate an extreme case. What is not shown is the fluctuation in power absorbed, since the friction varies through the engagement; this does not matter much in a mechanism such as the (reconstructed) London sundial-calendar, but it might in the Antikythera mechanism and certainly would be a cause for concern in designing a clock.

PRACTICAL RECONSTRUCTION

The rationale of reconstruction and the evidence for methods of manufacture gleaned from the original fragments have been described. Simple procedures for setting out and cutting toothed wheels have been discussed.

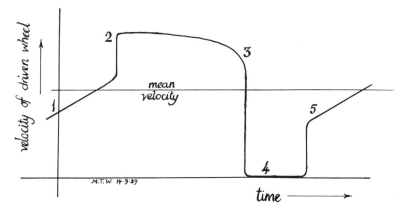

Figure 16 Graph to show variation of velocity ratio of the engagement shown in Figure 15.

This section is devoted to a record of the work of practical reconstruction.

Two complete instruments and a further set of wheels have been made to date (Figure 10). Modern materials that could be readily obtained were used. Most parts were cut from rolled brass sheet of 2.35 mm thickness (13 SWG); the vane of the sundial, ratchet wheel, day-of-the-week pointer and ring and shackle were cut from a thicker piece of similar material. The arbors were cut from drawn brass rod.

The strip to form the edge of the case was cut slightly short, bent up to form a butt joint and silver-soldered. It was stretched to size and made round by beating on a circular block, then silver-soldered to the front dial plate. In the first example some difficulty was experienced with distortion and it had to be gripped with tongs while hot to close the joint; after 'correction' with the hammer the case was neither round enough nor flat enough to be dressed in the lathe so it was filed up. The second example was chucked on a wooden disc and trimmed. Both were dressed by stoning with a block of pumice and, on the face to receive engraving, with a block of Water-of-Ayr stone (a fine slate), both used wet.

The lines on the front plate were laid out with compasses, straight edge and scriber. The holes were drilled. The lettering and heads for the days of the week were engraved by hand.

For both complete instruments the wheels were engine-cut. The rough discs were soft-soldered to the ends of pieces of brass bar to be chucked in a large ornamental-turning lathe (one having means of arresting the mandrel at determinate positions and of driving a rotating cutting instrument held at the toolpost). The discs were bored centrally and the edges cut into teeth using carbon steel flycutters (pieces of an old file) in a heavy universal cutting frame. The teeth were cut down until the root circle was of the required size. For wheels with nineteen to fifty-nine teeth, a 60° cutter was used; for the pinions of seven and ten teeth, cutters were made according to templates taken from an enlarged photograph of the originals. In each case the tip was rounded to match the shapes of spaces in the originals.

Unsoldered from their chucking pieces, the wheels were filed clean and the central holes opened to squares with a small square file. The radial lines and circles bounding the figures on the Moon disc and the circles marking out the large holes to be cut were scribed while this wheel was chucked for cutting. After dismounting from the bar it was chucked eccentrically on the faceplate for boring out the holes.

Both complete instruments were made to deadlines. The decision to use engine-cut wheels was prompted by the need to avoid uncertainties in the satisfactory outcome of the work. The clockmaker's technique was adopted, namely first making the wheels and then pitching them (adjusting their separation) to obtain good working. Had the action of the wheels proved defective, considerable work would have had to be redone, including making new arbors had the wheels needed removal; moreover, to use the same back dial plate it would have been necessary to make wheels to suit a pre-existing separation of the arbors which, with such wheels, was judged potentially awkward.

Rod for making the arbors was chucked in the lathe and turned to size.

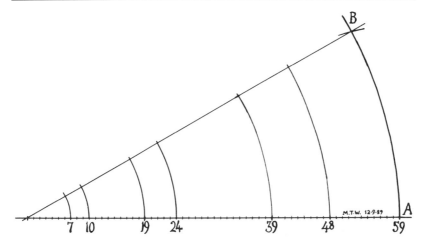

Figure 17 Sizing diagram for toothed wheels.

The square shanks onto which the wheels would be threaded were milled using a milling spindle at the toolpost and finished up to their shoulder by filing. They were parted off, the mobiles were assembled and the ends of the arbors riveted with the hammer.

One set of wheels, which has not been made up into a complete instrument, was made using only the hand techniques described in a previous section. The diagram of Figure 17 was laid out on a sheet of brass using compasses, straight edge and scriber. The horizontal scale is arbitrary; the radius of arc AB was taken from the surviving wheel of fifty-nine teeth. The diagram then gave the sizes for the root circles of all the wheels, which were scribed on the blanks, which in turn were filed approximately round. The blanks for the two wheels of fifty-nine teeth were soft-soldered together. Only the Moon disc was laid out, including the scribed circles bounding the numerals and circles for the holes that were drilled and filed out later, when the two wheels were separated.

For each of the wheels from nineteen to fifty-nine, the radius of the root circle was stepped off round the circle to divide it into six. The divisions for individual teeth were then stepped round by trial as described above. A separate small pair of compasses was kept for this task because once the opening had been adjusted for one wheel, it required only very slight adjustment for each of the others. (That it required adjustment at all is a consequence of the fact that the compasses measure chords, not arcs.) The two small pinions, of seven and ten teeth, were stepped out entirely by trial. The divisions were made more visible by ruling radially with a scriber and straight edge. The teeth were cut using a small saw file (a single-cut file of triangular or 'three-square' section) held in a comfortable handle (wider than is usual for a small file). The wheel to be cut was gripped in a leg vice, the leg of which was held in the vice of a joiner's bench, in order to bring the work to a high chest level for the writer, who worked standing. (Delicate filing is best done with the work much higher than the elbow height usually recommended for general work.) A notch was made with the corner of the file in each space between radial lines and was then deepened until it reached the root circle, while urging the file to right or left as necessary to keep the notch symmetrical between the radial lines. In general the eye was not aided although a weak (3 in) lens was occasionally used to check progress; the writer has long sight and did not then possess reading glasses. Several spaces could in general be filed before it was necessary to reset the wheel in the vice, but when the attempt was made to work too far from the topmost point it was found difficult to direct the file so that both sides of the notch made equal angles with the radius. Once one space was filed asymmetrically the tendency was to file the next one asymmetrically in the same sense. It was found easiest to work accurately if, in working at each space, either both adjacent spaces had been filed or both remained to be cut. Thus, imagining the spaces to be numbered consecutively, they were filed in a sequence such as 1, 3, 2, 5, 4, 7 . . .

For the wheels with nineteen teeth upwards, the file was merely sunk into the disc to produce each space. For the pinions of seven and ten teeth, in which the spaces were to be of a larger angle, the file was firstly sunk in as for the others but then each side of the space had to be worked on in turn. This would have made work on these small wheels rather slower anyway, but it was found particularly testing to produce symmetrical looking pinions, added to which it was a little tricky to handle them. Cutting the two wheels of fifty-nine teeth together called for careful work to ensure that the rear wheel was cut as evenly as the front one, but it was judged that even so the work was executed considerably more quickly than if they had been divided and cut individually.

The whole of this operation was carried out with reasonable but not extreme care, a conscious effort being made to do the work rather fast. In fact, averaging the whole set of wheels, and including both division (marking out) and cutting, the average time *per tooth* was about 30 seconds. Departures from regularity in the finished wheels were perceptible and out-of-roundness was measurable, but with care in pitching this set of wheels would work quite well. This was the first time that I had attempted this operation and there seems no reason to doubt that with practice and by taking greater care, wheels could be cut by hand to a quality only slightly inferior to the machine-cut ones.

These wheels were assembled onto arbors fashioned from rod entirely by filing, and the set was later mounted on to a disc of Perspex for display (Figure 10).

The numerals were engraved on the face of the Moon disc by hand, for which process the disc was held in a hollow in a block of wood. The face was then brightened by tinning: the disc was scoured clean and heated in a flame with sal ammoniac as a flux, and touched with a piece of grain tin; the molten tin was wiped thinly and evenly across the surface with a rag and the wheel was finally heated again with more sal ammoniac to smooth the surface by fusion. The openings were filled with hard black wax, cobblers' 'heel-ball'.

The rear dial was cut out to fit inside the rim of the case. Three small pieces were cut to be soft-soldered inside the case to support it at the right height, one of these having an upstanding nib engaging a notch in the rear dial to prevent it rotating, similar to the arrangement that registers the plates in an astrolabe.

The position of the day-of-the-week arbor was transferred from the hole in the front plate to the inside of the back plate. The position of the intermediate arbor between the two-lunation wheel and the year wheel, defined by a punch mark on the rear of the original front plate, was set out on both front and back plates. The mark on the front plate was used to position a collar, soft-soldered to the plate to act as a bearing for this mobile. The marks on the inside of the back plate were used as the basis of setting out the positions of all the mobiles on that plate. This was done by placing each pair of mobiles on the bench in engagement and taking the separation of their centres in the compasses which were then used to strike an arc from the centre marks already made. This procedure defined the centre for the lunation wheel and gave arcs for the loci of the centre of the two zodiac displays, which were further chosen to make these displays as large as possible. These latter two centres were drilled through with a small drill so that the zodiac rings could be scribed and divided on the outside. By further setting out with compasses, straight edge and scriber, the positions for the openings for the lunation display were marked. The smaller of these was drilled and filed to shape. The three large circular openings were cut in the lathe, mounting the plate on the faceplate. A hollow was cut in the plate to clear the rivet at the centre of the lunation disc. The plate was stoned clean and the abbreviated names of the zodiacal constellations were engraved by hand.

Three 'cocks' (brackets) were bent and filed up from brass sheet, one to carry the pivot of the lunation mobile, the others to hold the two zodiac display mobiles to their places. They were drilled, assembled with their mobiles on the rear plate and the rivet holes drilled through the plate. These holes were countersunk on the outside; rivets were made up, set by hammering and filed flush on the outside.

A horse-shaped pointer (like the *faras* of an astrolabe) was filed up to fit the cross-hole of the day-of-the-week arbor and a washer adjusted to fit under it allowing the mobile to turn freely without undue shake. A click-spring to work in the ratchet was cut from thin brass sheet, hammered to stiffen it, and adjusted in position to allow the pointer to point to the middle of each day. While it was clamped in position two rivet holes were drilled through the case and the spring, and countersunk. The spring was flush-riveted like the cocks.

The mechanism could now be put together and tried. In the first reconstruction the position of the collar soldered to the front plate was adjusted slightly to correct an error of pitching. The maker's name and date were engraved inside the case.

The balance of the assembly was tried on an edge so that the position and weight of the necessary counterpoise could be determined. A wooden cheek was clamped to the back of the case, which was stood on edge, and molten lead was cast in, slightly in excess. When cool it was pared down with a knife until a proper balance of the whole assembly was achieved.

The sundial vane was cut out. The stem was turned with a spigot to fit a small hole in the vane and these were wired together, with two pieces to form a boss, to be silver-soldered. The vane of the (small) Roman dial after which the reconstruction is styled is a one-piece casting; Tischendorf's sketch of the damaged vane of his example suggests a two-part construction.[29] The vane was filed up to balance about the axis of its stem, which was cross-drilled and slotted with a file for the cotter.

The suspension arm was bent up, filed to shape and stoned, the eye for the shackle having two tangs passing through squared holes in the arm and riveted on the inside. The ring was bent up from strip and silver-soldered (the original appears to have no joint) and was finished in the lathe (as the original seems to have been) on wood chucks. The shackle was bent and filed up and this assembly riveted together. A cotter was made for the stem of the vane, cotter and slot being adjusted to hold the arm stiffly in place. In the first reconstruction this cotter is quite plain; in the second it is again fancifully horse-shaped.

The whole instrument could now be assembled and its balance tried; in both instances it was found to hang with the dial plates very nearly vertical with no need to adjust the suspension. If the reconstruction is correct then the marked setting-over of the piece having the eye for the shackle in the original would be the result of accidental damage.

Made as described above, the first instrument occupied about 65 hours; the second about 70 hours. There is no obvious reason why the second took longer although it is the neater of the two articles. When it was made neither the original fragments nor the first reconstruction was readily available for comparison, both being on display.

Although an experienced tradesman might well work faster than even a skilled amateur, the workman of c. AD 500 would probably have been limited in speed by his tools. He would have had no hacksaw (but could have cut material with hammer and chisel on a bolster). His files and drills may well not have been as efficient, and very probably his workholding devices were not as convenient as the screw vices used. In particular, if he used a lathe at all it would undoubtedly have been less convenient than the one used in reconstruction. Yet it should be clear that all the work done (for convenience) in the lathe could quite well have been done by hand; usually it would take longer and sometimes the work would be detectably less regular. It should also be ·remembered that the reconstructions were made from prepared brass plate and rod; we do not know whether the maker of the original could obtain such 'stock' or had to prepare his own. Yet the experience of practical reconstruction, as well as showing that the reconstruction (on paper) is workable, demonstrates very clearly that such an instrument could be made with simple tools, moderate skill, and in a reasonable length of time.

AL-BIRUNI'S 'BOX OF THE MOON' RECONSIDERED

It was argued above that al-Bīrūnī's wheel of forty should, in our reconstruction, be replaced by one of thirty-nine. If this point suggests the possibility of a corruption of al-Bīrūnī's text, then it is a corruption that exists in all known manuscript sources; on the available evidence it seems most likely that forty was the number given by al-Bīrūnī. But there are other indications that al-Bīrūnī did not devise the 'Box of the Moon' and that he was reporting a well-established, even a familiar, device.

With his known mathematical ability and with the astronomical data available to him, al-Bīrūnī should have been aware of the errors of the instrument and would have been quite capable of realizing that substituting a wheel of thirty-nine for that of forty would have effected a great improvement. Had he himself devised the instrument it seems most probable that he would have specified the wheel of thirty-nine. A man who could specify two wheels of fifty-nine would surely not be shy of one of thirty-nine, and the workman who could make the other wheels would have no difficulty with the thirty-nine, even if it took a little longer to divide than a wheel of forty.

The description of the 'Box of the Moon' comes at the end of a text on the astrolabe and the device is introduced in a matter of fact way, almost casually.[30] There is no suggestion that it is novel and there is no description of what it is for; the reader should presumably know that. This is in contrast to the detailed description of how it is to be made.

The sizes of the wheels are given in relation to the size of the box, and here as with the wheel of forty we have a detail that does not make sense. The box is far too large, or the wheels are much smaller than could be fitted. Enlarging the box in this way does not allow the maker to create a more handsome display, because the size of the dials is limited by their separation, which depends on the sizes of the wheels. The proportions given merely make the instrument more cumbersome than it need be or more difficult to make than it need be (because the wheels are smaller than they could be). Again I speculate that al-Bīrūnī may have been reporting a corrupt tradition.

In describing how the toothed wheels are to be sized, al-Bīrūnī appears to state that the radii of the circles bounding the *tips* of the teeth should be in proportion to the number of teeth. This procedure leads to wheels of which the smaller are, relatively speaking, undersized, and the effect is serious with the extreme case that occurs in the instrument, i.e. of a pinion of seven leading a wheel of fifty-nine. Although he seems rather fussy about how the teeth are to be formed, these do not seem to be the instructions of a man with much first-hand experience of making the mechanism.

Finally, al-Bīrūnī goes on to report some other, less adequate, trains of wheels to work similar displays. Here he makes it plain that he is reporting the work of others.

In all, it seems as though al-Bīrūnī was reporting a device that was, in general terms, well known. It appears that he thought it sufficiently interesting to his readers to present a particular description of how to make it but it may be doubted whether he had actually done the work himself. Its use was either well known or it was regarded as a useless curiosity, perhaps a traditional instrument whose use had been forgotten. It does seem possible that he was reporting a tradition that could have become corrupted in at least some details.

BEYOND THE MINIMAL RECONSTRUCTION

In the reconstruction outlined above, the wheelwork is completed according to a train similar to al-Bīrūnī's so that, besides the indications of day of the week and age and phase of the Moon, there are displays of the position of Sun and Moon in the zodiac (and possibly date in the Julian calendar). This is the simplest reconstruction that I have been able to devise which accounts for all the features of the surviving fragments. As Field and I have noted,[31] we cannot with any confidence identify a coherent purpose or set of purposes for the ensemble presented by this calendrical device built into an otherwise conventional sundial, but the personal reactions of those who have handled the practical reconstruction have reinforced our view that it is a satisfying gadget and perhaps this is a sufficient rationale.

Even so it is possible to think of elaborations of this scheme which could equally accord with the evidence of the surviving fragments, some of which, if developed, might suggest uses for the instrument as a whole.

In the earlier paper[32] we pointed out that it is possible for the Sun-in-the-zodiac and Moon-in-the-zodiac indications to be brought together in one concentric display, as in the later Oxford instrument. No further information is gained thereby. This can only be done by the rather unpleasing extension of one or both ends of the train by idle wheels. It appears that the Antikythera mechanism incorporated such a concentric display but I do not feel justified in introducing the elaboration here.

Eclipse Predictor

The calendrical mechanism could be used to predict eclipses if one knew the position of the Moon's *nodes*. These are the points in the zodiac at which the (apparent) paths of the Moon and Sun intersect. They are at diametrically opposed points which move slowly against the background of stars in a direction opposite to that of the Sun and Moon, making a complete circuit in approximately 18.6 years. Eclipses of the Moon take place at Full Moon and those of the Sun at New Moon, but only when they are near to the nodes (that is, when they and the Earth are nearly collinear). The positions of the nodes could be shown quite simply on one or the other of the zodiac dials. Such an indication added to the Moon's zodiac scale could balance visually a Julian calendar ring concentric with the Sun's zodiac scale, as suggested above. The nodes could be indicated by pegs in a series of holes around the dial. A precedent for this system is found in fragmentary dial plates for anaphoric clocks dating from the first to third centuries AD (in which the position of the Sun in the zodiac was indicated in this way).[33] Equally the nodes could be represented by marks on a ring fitted friction-tight to the dial plate and again moved by hand.[34] But it is possible to envisage such a ring being moved automatically. The logical starting point is the Sun-

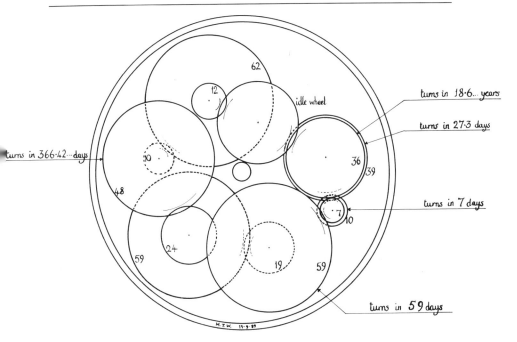

Figure 18 Conjectural reconstruction of wheel train for display showing motion of Moon's nodes.

in-the-zodiac (one-year) wheel, the slowest-moving part of the minimal reconstruction.

Two possible trains will suffice as examples:

$$62/10 \times 36/12 \text{ (output turns in 18.6 years)}$$
$$59/10 \times 38/12 \text{ (output turns in 18.68 years)}$$

An additional idle wheel would have to be included in each case to reverse the motion; either train could be fitted within the instrument, as shown by Figure 18. Figure 19 shows how the concentric mobiles for the Moon and for

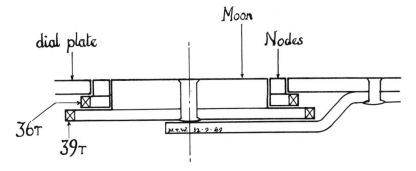

Figure 19 Display of Moon's nodes: a possible arrangement viewed in section.

Figure 20 Display of Moon's nodes: appearance of display shown in section in Figure 19.

its nodes might be arranged, and Figure 20 shows how the dial might appear. The attraction of the second, less accurate, train lies in its use of a wheel of fifty-nine teeth, a number appearing in the original fragments, and perhaps the inclusion of 38, which is twice 19—a number also used in the original fragments. The first train may be made more attractive in these respects either by adopting the wheel pair 48/16 for 36/12 (if we suppose the original contained 48 as in al-Bīrūnī's train) or 39/13 (if we suppose the original contained 39 as in the 'improvement' to al-Bīrūnī's train). Both trains, as given above, contain 10, which appears in the original fragments, and 12, a factor of 48.

Use of the Dial as a Moon Dial

We have previously outlined how the calendrical mechanism might have facilitated telling the time by the Moon.[35] This function also would suggest some elaboration of the minimal reconstruction.

The vane of the dial would have to be set in accordance with the Moon's zenith distance when in the observer's meridian. The main component of this adjustment, the declination of that point on the ecliptic having the same celestial longitude as the Moon, would in effect be given by the display showing the Moon's position in the zodiac. This would have to be converted to the date at which the Sun is at that point for setting the vane. One might imagine the Moon's zodiac scale also (or alternatively) engraved with the months, although it would seem less 'Byzantine' to have engraved the signs of the zodiac on the declination scale of the dial. The elevation of the vane would have also to be adjusted for the Moon's celestial latitude, which is determined by the longitude between it and its nodes. The Moon's latitude could therefore be found with the aid of a display showing the position of the nodes such as that suggested above. The Moon's latitude is not in general to be added directly to its declination. A graphical correction on the declination scale of the dial is possible, but since no such extra scale is present on the surviving dial one must suppose the Moon's latitude to be multiplied by the appropriate factor, which varies between cos 23½° and 1 (but since the latitude is always small it might have been taken as unity at all times). This could have been performed by use of a table or graphically on a diagram, either of which could have been laid out on the back of the instrument.

Having set the dial and obtained a reading, the user would then have to convert the Moon hours thus found to the temporal hour system, which was presumably used since it is to this system that time shown on the dial by the Sun approximates. It would be necessary to allow for the difference in right ascension or hour angle between Sun and Moon; this is related to their difference in celestial longitude which is given by the 'age-of-the-Moon' display or by comparison of the 'Sun-in-the-zodiac' and 'Moon-in-the-zodiac' displays. As with the conversion from celestial latitude to declination discussed above, a factor between cos 23½° and 1 is involved so that a single table or graphical representation might serve for both conversions.

In principle another type of conversion would also be necessary: the lengths of the hours shown by Sun and Moon depend on their respective declinations and the (geographical) latitude at which the dial is used. Again the declinations could be taken from the back dial display, but a table or diagram for this computation would necessarily be more complex than that envisaged above because of the dependence on latitude; for example, one might use a diagram comprising a set of curves for different latitudes.

Automatic Setting of the Sundial Vane

As was described in the earlier paper[36] the vane of the sundial must be set according to the declination of the Sun for the time of the year, and a scale of months is accordingly engraved on the dial plate. It has been mentioned that the Sun-in-the-zodiac display might also, or instead, be engraved with the names of the months to facilitate this setting. Zeeman has suggested that by the addition of a simple mechanism linking this wheel to the vane, the vane might be set automatically.[37] The declination of the Sun varies approximately with simple harmonic motion through the year (the Sun's declination actually varies sinusoidally with its right ascension or sidereal hour angle whereas it is the Sun's celestial longitude which changes uniformly with time) so that a crank pin set in the wheel that rotates once in a year could be made to yield an appropriate motion. The obvious mechanism, a slotted oscillating lever acting on the pin, introduces a further approximation which may, however, be reduced if instead of working the vane directly this lever acts on a second one attached to the vane. It seems rather unlikely that any scheme yielding a more exact motion could be made plausibly in character with the rest of the instrument. The presence of any such mechanism would preclude the convenient setting of the dial for use as a Moon dial as described above. It would also make it unnecessary to have the the scale of months engraved on the dial as in the surviving fragments. Therefore, while it is an intriguing idea, I think it most unlikely that such a mechanism was fitted.

CONCLUSIONS

The original fragments of the London sundial-calendar provide sufficient evidence for the general form of the instrument to be reconstructed with considerable confidence, and yield evidence of the use of a repertoire of easily attainable metalworking techniques which has been found sufficient in making practical reconstructions.

Applying the concept of the 'minimal reconstruction' to the problem of the missing wheelwork results in a mechanism closely parallel to that described by al-Bīrūnī as the 'Box of the Moon'. One departure from al-Bīrūnī's scheme, the substitution of a wheel of thirty-nine teeth for one of forty in the display of the Moon's position in the zodiac, has been made in building practical reconstructions, and the argument has been advanced that this and other details of the device as reported by al-Bīrūnī may have become corrupted, probably at a date prior to the writing of the text in which al-Bīrūnī seems to report a traditional device.

It has been pointed out that by a further departure from al-Bīrūnī's scheme the calendrical mechanism could be reconstructed to embody the mathematically excellent relationship between lunation and year known as the Metonic ratio. Other variations to the wheelwork, also giving acceptable calendrical ratios, are equally compatible with the evidence of the surviving fragments, but the general nature of the calendrical displays as reconstructed 'minimally' remains unchanged.

The exploration of these possibilities in no way invalidates the earlier conclusion that the London sundial-calendar provides strong evidence for the transmission of a tradition of mathematical gearing from Hellenistic to Islamic culture via the Byzantine world.[38]

Discussion of the problems of designing and making toothed wheels leads to the conclusion that the difficulties that have been supposed to exist in these operations for the workmen of antiquity are illusory. While we do not know what methods and tools these men used, and while such tools as division plates are certainly a convenience, practical trial has demonstrated that toothed wheels like those found in the London sundial-calendar can easily be made to a good standard and quite quickly, by the very simplest means.

The experience of making practical reconstructions of the London sundial-calendar has demonstrated that the entire instrument could be made with a modest repertoire of techniques and simple tools in a reasonable period of time. The skill and labour required were not prohibitive, and since the finished instruments each weigh only just over 1 kg the cost of materials, compared to that required for, say, a metal helmet, does not appear prohibitive either. Thus there seems no reason to suppose that the instrument was necessarily particularly rare. This deduction is compatible with the observed quality of the surviving fragments.

Some possible elaborations of the minimal reconstruction have been considered because, although they have no historical basis whatever, such further reconstruction could provide a link between the two apparently independent parts of the device. Such a link could be a physical connection or, more probably, a logical one in which the two functions, still physically independent, were applied to a common purpose. However, unless further evidence, such as literary references or fragments of a comparable instrument, comes to light it is unlikely that such speculative exploration can ever be conclusive. Yet the effort to find a unifying purpose seems worth making in that we are here dealing with the earliest known combination instrument.

In any case, the practical reconstructions have demonstrated to those who have handled them that, even without such a unifying purpose, the instrument is a delightful and covetable gadget. Perhaps we should no more expect to find a single purpose for the parts than we should for the several blades of a Swiss Army penknife, except that overriding function of getting money out of a gentleman's pocket. In other words, the explanation for the instrument may perhaps be found in the mentality of its potential purchaser. Identification of a purpose would enhance our understanding and appreciation of the instrument but the lack of it does not diminish the historical significance of the fragments.

Acknowledgements

The origins of this paper are to be found in the work done jointly on the London sundial-calendar in 1983 and 1984 by my colleague Dr J.V. Field and myself. As it has developed I have benefited from discussion with Dr Field and with other colleagues and friends who have taken an interest in the instrument. Among these I am particularly grateful to Professor A.G. Bromley (University of Sydney; also a visiting fellow of the Science Museum), Dr D.R. Hill, my former colleague Mr R.J. Law, Commander J.D. Richard and Professor C. Zeeman (Hertford College, Oxford). I have to thank my wife for her forbearance while the reconstructions were made and her greater forbearance while the paper was drafted, and my sons for conducting user field trials of the instruments at their boarding school. I thank Professor Bromley and Dr Field for helpful comments on an early draft and Ms D. Hunter for her typing skill. Any remaining errors are my own. Photographs are reproduced by courtesy of the Trustees, The Science Museum; negative numbers as follow: Figure 1a, 953/83/6; 1b; 953/83/8; 2a, 118/89; 2b, 117/89; 2c, 1124/84/1; 3a, 955/83/12; 3b, 955/83/9; 4a, 954/83/4; 4b, 954/83/6; 8, 953/83/1; 9, 1124/84/7; 10, 154/89; 11a, 150/89; 11b, 151/89; 12, 153/89; 13, 152/89.

References

1. Inventory Number 1983-1393.

2. J.V. Field and M.T. Wright, 'Gears from the Byzantines: a portable sundial with calendrical gearing', *Annals of Science*, Vol. 42, 1985, pp. 87-138. Reprinted in J.V. Field, D.R. Hill and M.T. Wright, *Byzantine and Arabic Mathematical Gearing*, London, The Science Museum, 1985.

3. D.J. de S. Price, 'Gears from the Greeks', *Transactions of the American Philosophical Society*, New Series Vol. 64, 1974, part 7.

4. A.G. Bromley, 'Notes on the Antikythera mechanism', *Centaurus*, Vol. 29, 1986, pp. 5-27.

5. D.R. Hill, 'Al-Bīrūnī's mechanical calendar', *Annals of Science*, Vol. 42, 1985, pp. 139-163; also reprinted in Field, Hill and Wright, op. cit., see reference 2.

6. See, for example, V. Foley, W. Soedel, J. Turner and B. Wilhoite, 'The origin of gearing', *History of Technology*, Vol. 7, 1982, pp. 101-129.
7. Field and Wright, op. cit.
8. C. Holtzapffel, *Turning and Mechanical Manipulation*, Vol. 1, 1843.
9. Cf. the casting of blanks for coining: see, for example, R.F. Tylecote, *Metallurgy in Archaeology*, 1976.
10. See, for example, Tylecote, op. cit., reference 9.
11. Price, op. cit.
12. Hill, op. cit.
13. Field and Wright, op. cit.
14. Field and Wright, op. cit.
15. Field and Wright, op. cit.
16. Field and Wright, op. cit.
17. Hill, op. cit.
18. J.V. Field and M.T. Wright, *Early Gearing*, The Science Museum, London 1985.
19. Price, op. cit.
20. This elegant and economical arrangement was suggested by Mr R.J. Law, late of the Science Museum: private communication.
21. Hill, op. cit.
22. (pseudo-) Aristotle, *Opuscula*, Mechanica; see, for example, W.D. Ross (ed.), *The Works of Aristotle*, Vol. VI, 1913.
23. J. Smeaton, 'Observations on the graduation of astronomical instruments; with an explanation of the method invented by the late Mr Henry Hindley, of York, clock-maker, to divide circles into any given number of parts', *Phil. Trans.*, Vol. LXXVI, 1786.
24. Price, op. cit.
25. A.G. Bromley and F.A. Percival, *Methods of Dividing Gearwheels Accessible to the Ancients*, Technical Report 296, Basser Department of Computer Science, The University of Sydney, NSW, Australia, 1988.
26. A. Brieux, Histoire des Sciences, *Livres—Instruments—Autographes Novembre 1977*, [catalogue] 1977.
27. Price, op. cit.
28. Price, op. cit.
29. L.F.C. Tischendorf, *Notitia editionis codicis bibliorum siniatici* . . . [Leipzig, 1860]; the figure is reproduced by E. Buchner, 'Antike Reiseuhren', *Chiron*, Vol. 1, 1971, 457-482 (Figure 7).
30. Hill, op. cit.
31. Field and Wright, op. cit., reference 2.
32. Field and Wright, op. cit., reference 2.
33. O. Neugebauer, *A History of Ancient Mathematical Astronomy*, Springer-Verlag, Berlin and New York, 1975, p. 870.
34. Cf. the calendar ring on the Antikythera mechanism reported by Price, op. cit.
35. Field and Wright, op. cit., reference 2.
36. Field and Wright, op. cit., reference 2.
37. C. Zeeman, private communication.
38. Field and Wright, op. cit., reference 2.

Some Roman and Byzantine Portable Sundials and the London Sundial-Calendar

J.V. FIELD

In 1983 the Science Museum, London, acquired four fragments of a portable sundial with associated calendrical gearing. These fragments are shown in Figure 1. They are a circular plate, two arbors which between them carry four gear wheels and a ratchet, and a suspension arm. M.T. Wright and I have described elsewhere the various investigations we carried out to assure ourselves, as far as we could, that the four fragments, which came to us without precise provenance, might indeed be parts of the same instru-

Figure 1 The four fragments of the London sundial-calendar. Collection Science Museum, London, Inv 1983-1393. (ScM neg. no 955/83/3) (*Reproduced by permission of the Trustees of the Science Museum, London*)

103

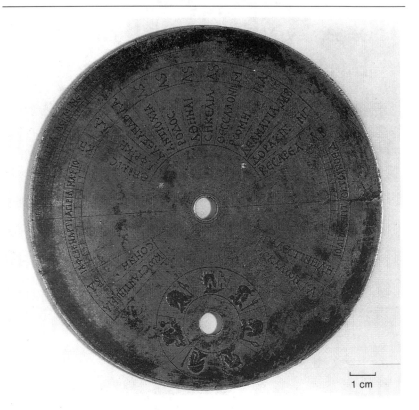

Figure 2 The sundial plate of the London sundial-calendar. Collection Science Museum, London, Inv 1983-1393. (ScM neg. no 953/83/6) (*Reproduced by permission of the Trustees of the Science Museum*)

ment and could have been made in the early Byzantine period (AD 328–641).[1] The large round plate (shown separately in Figure 2), which is marked out as a sundial, could be dated to this period on stylistic grounds. A similar dating was suggested by considering the names of the places included in the list of latitudes inscribed in the space around the sundial markings.

It was of particular importance to establish as precise a date as possible for our fragments since it was clear that two of them, namely the arbor carrying a Moon disc with fifty-nine teeth and a pinion of nineteen and the arbor carrying gears of seven and ten teeth, corresponded with two parts of a mechanical calendar described by the Persian scholar al-Bīrūnī in about AD 1000 (Figure 3).[2] The Moon disc even appeared to have been laid out exactly as al-Biruni described. The only difference between the Byzantine fragments and the corresponding parts of the Islamic instrument was that the Byzantine arbor with gears of seven and ten also carried a seven-lobed ratchet.

Al-Bīrūnī's instrument, which he calls 'the Box of the Moon', is designed to show the shape of the Moon (roughly), its age in days and the positions of the Sun and the Moon in the zodiac. We quickly found that gear trains exactly like those in the 'Box of the Moon' could be fitted within the space available in the Byzantine instrument—the space being defined by the size of the large disc and the thickness enclosed by the suspension arm (Figure 4).

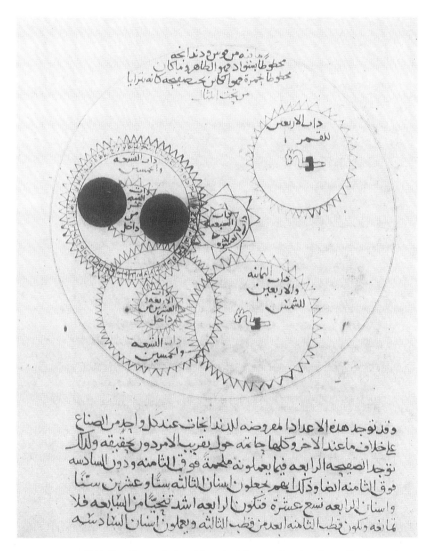

Figure 3 Al-Bīrūnī's 'Box of the Moon'. Manuscript in the collection of the University Library, Leiden (MS Or 591(4)). (*Reproduced by permission of Leiden University Library*)

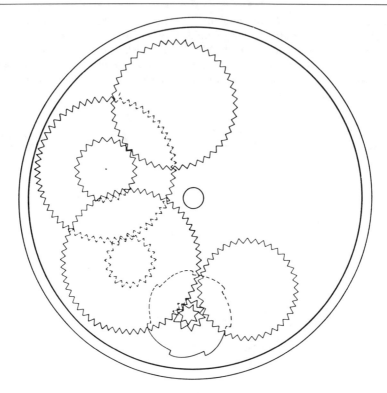

Figure 4 The calendrical gear trains described by al-Biruni arranged so as to fit behind the sundial plate of the London sundial-calendar. (ScM neg. no 575/85) (*Reproduced by permission of the Trustees of the Science Museum*)

Moreover, consideration of the possible outputs from trains leading on from the gears of the surviving arbors, and fitting within the available space, made it clear that the Byzantine instrument could not have been very different from that described by al-Biruni, though there may have been some slight differences. The gearing of the London sundial-calendar thus provides a Byzantine precedent for the device described by al-Biruni. In fact, al-Biruni himself seems to regard the device as 'traditional' since he gives no inventor for it, whereas he does give inventors for other devices.

As we shall see below, the Islamic adoption of this Byzantine calendrical device may be of interest in connection with considerations of the possible usefulness of our sundial-calendar as a whole. The Byzantine device is, however, of interest in its own right as providing evidence that the Hellenistic tradition of making mathematical gearing, attested by the mechanism recovered from the Antikythera wreck (datable to the first century BC), continued to be active in the early Byzantine period. As far as one can tell, given the very different states of preservation of the objects to be compared, the Byzantine gear wheels seem to be very similar to those of the

Antikythera mechanism. The Byzantine teeth are somewhat larger, but have the same 60° shape, no doubt as a result of their being formed by means of a 60° file of the kind used, for example, to sharpen saws. (This suggestion was first made to me by M.T. Wright.) Moreover, both sets of gears seem to have been fitted on square arbors. It nonetheless seems clear that the Byzantine gear trains are not sub-components of the much more elaborate trains found in the Antikythera device, so the later mechanism is not to be interpreted as a debased version of the earlier one. Indeed, if we are correct in supposing that one of the outputs of the Byzantine gear trains was the position of the Sun in the zodiac, then it seems likely that some such form of mechanical calendar was invented before the Julian calendar reform of 38 BC. For once one is using the Julian calendar the position of the Sun in the zodiac is given by the date. While one may imagine that a train of gears continued to be employed after it could have been replaced by a simple list of dates and positions, it seems rather unlikely that such a train would have been invented *de novo* at such a stage. However, since so little is known of Byzantine technology, rational reconstruction is a more than usually hazardous process and we cannot rule out the possibility that our gear train presents a mechanical counterpart to the intricate political manoeuvrings for which the Byzantine Empire has become a byword.

The gearing of the London sundial-calendar is, so far, unique. Nonetheless, as has been explained in more detail elsewhere, we have no reason to suppose it was a very extraordinary object in its own day.[3] Work on the sundial part of the instrument was originally undertaken as a means of dating the calendrical device. It has, however, proved to be useful as a means of situating the instrument as a whole in some kind of historical context, for it seems that the character of the instrument as a whole is rather in accordance with what we know of the sundial part, namely that it was quite common. Of the seven other Byzantine 'scientific' instruments known, no less than four are sundials of exactly this type.[4]

OTHER DIALS OF THE SAME TYPE

The largest of the four surviving fragments of the London sundial-calendar is the circular plate shown in Figure 2. The sundial markings are (i) a quadrant scale on the rim, running clockwise from zero to 90°, numbered in intervals of 5° but actually divided into single degrees (these finer divisions apparently being put in by eye); and (ii) the double fan (labrys-shaped) pair of scales marked with abbreviated forms of the Greek versions of the months of the Julian calendar. Apart from the four months closest to the solstices, whose space on the scales is rather short, all the months are subdivided into thirds (corresponding to the division of zodiac signs into decans). A possible method of setting out this pair of scales has been thoroughly described by Buchner.[5] It involves using the 'analemma' construction given by Vitruvius. However, the subdivisions of the months on the London dial look rather too nearly equal to one another and may have been put in by eye.

As we have already seen, this dial is of a well-known type, there being four other known dials of closely similar design with inscriptions in Greek. Two

108 Some Roman and Byzantine Portable Sundials

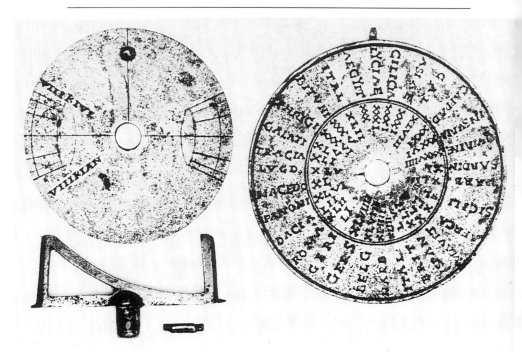

Figure 5 Roman portable sundial, disassembled. Collection Museum of the History of Science, Oxford, LEI 1. (*Reproduced by permission of the Museum of the History of Science, Oxford*)

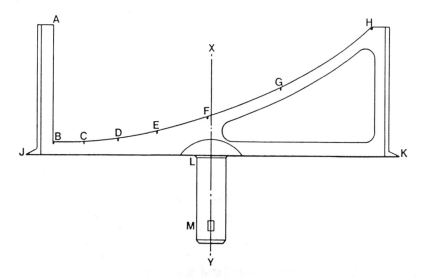

Figure 6 The combined gnomon and hour scale of the London sundial-calendar (reconstruction). (*Reproduced by permission of the Trustees of the Science Museum*)

are the product of recent, well-recorded archaeological excavations: at Aphrodisias in Asia Minor[6] and on the island of Samos.[7] Another was found in the necropolis at Memphis and published by the Biblical scholar Tischendorf in 1860—as an 'astrolabium'.[8] The fourth appeared in the sale catalogue of the Parisian instrument dealer Alain Brieux in 1977 and was purchased by the Time Museum (Rockford, Illinois, USA). All four of these dials have been dated to the early Byzantine period. We shall discuss below some possible indications of how they may be put into chronological order within this period.

There are also three known examples of portable dials of this type with inscriptions in Latin. One was discovered near Rome in 1740 and described by Baldini in a paper read to the Accademia Etrusca in the following year.[9] A second was discovered on the site of a Roman settlement at Crêt-Chatelard (Loire) in about 1837. It was described with exemplary clarity in a paper by Durand and La Noë which was read to the Société nationale des Antiquaires de France in 1897.[10] Unfortunately, the relevant publications now seem to be all that remains of these two dials. The published illustrations of the dials have been included in the present paper (see below, Figs. 9 to 11) since the original publications are no longer easily obtained. The third dial of this type with an inscription in Latin is said to have been found near Bratislava early in the present century. It was acquired shortly thereafter by Lewis Evans and is now in the collection of the Museum of the History of Science, Oxford (inventory number LEI 1).[11] As a museum exhibit it has the splendid advantage of being complete. It is, in fact, the only known complete example of this type of dial and thus provides crucial evidence as to how these dials worked.

HOW THE DIALS WORKED

It is obvious that the Science Museum's dial, shown in Figures 1 and 2, lacks both a gnomon and a scale upon which its shadow is to be cast. In fact, we can see from the Oxford dial, shown disassembled in Figure 5, that gnomon and scale were combined into a single piece (bottom left in Figure 5). This is shown diagrammatically in Figure 6. The point A is the tip of the gnomon, whose shadow is cast on the curved scale BH. The stem LM passes through the dial plate so that the face JK lies against it. Pointers at J and K register against the scale of months on the dial. The line AB is perpendicular to the plate. The line AH is parallel to JK, and thus parallel to the dial plate. It marks the direction of the ray of the Sun at noon. The hour markings at B, C, D, E, F and G are placed so that the intervals BC, CD, etc. each subtend equal angles (of 15°) at A. This hour scale will clearly be non-linear unless the arc BH is part of a circle centre A. The hour scale on the Oxford dial is severely non-linear, so that it would be difficult to guess at subdivisions of an hour. In fact, subdivisions of the hour are not mentioned in what survives of Ancient literature and they do not seem to have been employed on Ancient devices for measuring time. The astronomer used minutes and seconds in his work, but for the layman time was given as at some particular hour or between two hours.

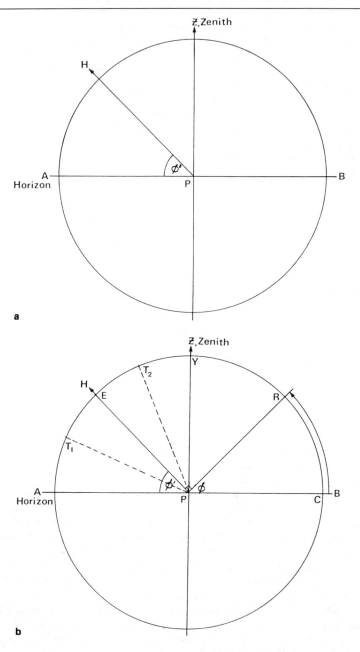

Figure 7 Layout of scales on portable sundials. (a) Meridian plane, showing height of Sun (H) at noon at the equinox; (b) face of dial in meridian plane, showing adjustment using anticlockwise scale for latitude; (c) (opposite) face of dial in meridian plane, showing adjustment using clockwise scale for latitude. (*Reproduced by permission of the Trustees of the Science Museum*)

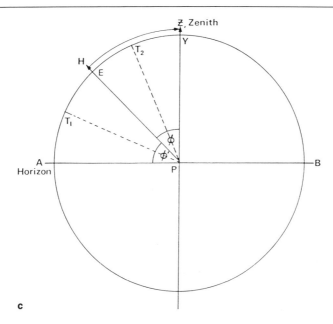

c

The principle of how this type of dial worked has been described in detail by Stebbins, who shows that the design involves an ingenious approximation that may introduce errors of up to about a quarter to an hour.[12] This error is, of course, of more interest to the modern investigator than it would have been to a Roman or Byzantine user, even supposing he had the means or the curiosity to check the reading of his dial against that of another sundial. The expected outcome of such curiosity is suggested by Seneca's well-known comment that philosophers agree more often than clocks.[13]

As can be seen from the design of the combined gnomon and scale, the functioning of the dial depends upon the baseline of this part (JK in Fig. 6) pointing directly at the Sun at noon. The shadow of the gnomon will then reach the top of the scale as it should.

Let us first consider the simplest case, when the Sun is on the equator, i.e. the moment of the equinox, which we shall take to occur at noon. The altitude of the Sun above the horizon is then equal to the observer's co-latitude, i.e. it is $90° - \phi$, where ϕ is the latitude at which the observation is made.[14] We shall adopt the usual convention of calling the co-latitude ϕ'.

Figure 7a shows the meridian plane through the observer, which must clearly include the face of the dial, whose centre is given as P. Let the observer's zenith be Z and his horizon AB. The Sun then lies in the direction PH, where $\angle HPA = \phi'$. To use the dial, we require a latitude scale that allows us to set the baseline of the gnomon/scale along the line PH.

One possible way of achieving this is shown in Figure 7b. Here we have drawn two radii, PR, PE, of the dial, making right-angles with one another. By working through the angles around the point P we can see that in order to make PE lie along PH we must set the line PR so that $\angle RPB = \phi$, the latitude

Figure 8 Roman portable sundial, assembled. Collection Museum of the History of Science, Oxford, LEI 1. (*Reproduced by permission of the Museum of the History of Science, Oxford*)

of the observer. There is clearly no problem in setting some line of the dial, say PY, to be vertical, since it will suffice for the dial to be suspended from the point Y. Then if we inscribe a latitude scale anticlockwise in the first quadrant of the dial, that is, along the arc CY, we can make the adjustment of the pair of perpendicular radii, PR, PE, by moving PR against this scale until $\angle RPC = \phi$. This design has given us a latitude scale fixed with respect to the point of suspension of the dial (i.e. the point Y) and a line to show the position of the baseline of the gnomon/scale that turns with respect to the latitude scale. It is clear that the further adjustment necessary to allow for the time of year, i.e. for the distance of the Sun from the equator (its declination), can be made by means of declination scales that will form a fan shape extending $23\frac{1}{2}°$ (in principle) on either side of PE. The edges of the fan are shown by dashed lines in Figure 7b and the end points of its arc, which correspond to the solstices, have been marked T_1 and T_2.

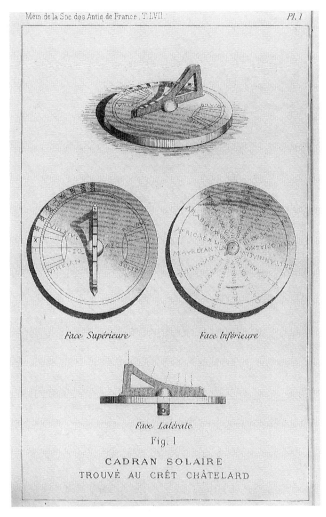

Figure 9 Roman portable sundial found at Crêt-Chatelard. First plate from Durand and La Noë, 'Cadran solaire portatif . . .' (note 10). (ScM neg. no 1048/85) (*Reproduced by permission of the Trustees of the Science Museum*)

DIALS WITH INSCRIPTIONS IN LATIN

The markings on the Roman dial now in the collection of the Museum of the History of Science, Oxford, correspond exactly to the design we have just described. Figure 8 shows the dial assembled. The small ring on the rim at the top of the disc is presumably intended to take a suspension cord. It corresponds to the point Y in Figure 7b. The first quadrant of the rim of the dial carries a latitude scale, running anticlockwise (markings are shown only for latitudes from 30° to 60°, corresponding to the range in latitudes of the places listed on the back of the dial, shown at the right in Figure 5). The

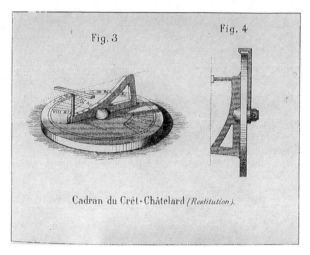

Figure 10 Roman portable sundial found at Crêt-Chatelard. Detail of second plate from Durand and La Noë, 'Cadran solaire portatif . . .' (note 10). (ScM neg. no 1049/85) (*Reproduced by permission of the Trustees of the Science Museum*)

smaller disc, shown at the left in Figure 5, carries markings corresponding to the lines shown in Figure 7b as PR, PE, PT_1, PT_2. There are also additional lines, corresponding to divisions between the signs of the zodiac, and the pattern of lines has been made symmetrical about the centre of the disc. When the dial is assembled, as shown in Figure 8, this smaller disc is free to turn within a depression in the larger one, so that the position of the line corresponding to PR in Figure 7b can be set against the latitude scale. A small knob is supplied on this line for the user's convenience (and possibly also as a reminder of how the instrument is to be set?). Figure 8 shows the dial set, somewhat unconvincingly, for use at about at about 88°N, some way beyond even *ultima Thule*.

The Oxford dial is thus designed so that the latitude is set up on a scale fixed with respect to the point of suspension and running anticlockwise in the first quadrant. This adjustment then sets up a perpendicular line in the direction of the equinox. That is, lines corresponding to PR and PE in Figure 7b move together with respect to the latitude scale. If we turn to the dial from Crêt-Chatelard, shown in Figures 9 and 10, we see that its form of adjustment seems to have been different from that of the Oxford dial, since now the latitude scale runs clockwise and apparently starts exactly at the line marking the equinox. It is, of course, possible that this latter characteristic is due to an accident of preservation which caused two originally separate parts to be cemented together by corrosion in this particular relationship. We may note, however, that there seems to be no line to correspond to that shown as PR in Figure 7b, and no knob like that on the corresponding line in the Oxford dial. There is, in fact, no reason why we should not accept the more straightforward explanation of the markings shown in Durand and La Noë's drawings of the Crêt-Chatelard dial, namely that they indicate that the dial

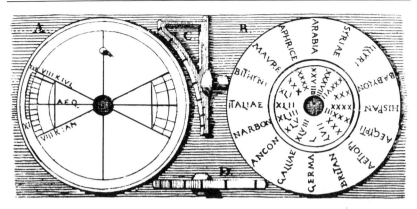

Figure 11 Roman portable sundial found near Rome. Figures from Baldini, 'Sopra un'antica piastra di bronzo . . .' (note 9). (*Reproduced by permission of the Museum of the History of Science, Oxford*)

was designed to be adjusted in a manner slightly different from that employed for the Oxford dial. As we shall see, other dials are known to employ the same form of adjustment as that seemingly implied by the drawings of the Crêt-Chatelard dial..

This form of adjustment is shown diagrammatically in Figure 7c. It depends upon setting the angle HPZ to be equal to the latitude ϕ of the observer, thus making '$HPA = \phi'$, his co-latitude. It is clear that this adjustment can be made by moving the point of suspension, Y in Figure 7c, against a clockwise scale of latitude starting from the line PE. We thus have a latitude scale fixed with respect to the equinox line and a point of suspension which is movable. An economical method of effecting such an adjustment would be to have a suspension arm that pivoted about the centre of the disc. This is the arrangement suggested by Durand and La Noë, and shown in one of their illustrations, reproduced in our Figure 10. (The only dial for which the suspension arm actually survives is the London sundial-calendar—see below and Figures 1 and 12.)

The dial found near Rome in 1740 seems to have resembled the Oxford dial rather than the one found at Crêt-Chatelard. Baldini's drawings, reproduced in our Figure 11, show the pair of perpendicular lines through the centre of the dial and the small knob on one of them that was presumably to be used in moving the inner plate. However, the drawings, which correspond closely to the accompanying text, also present certain puzzling features such as the apparently asymmetrical markings of the declination scales. These features may perhaps be due to the difficulty encountered in interpreting marks on a heavily corroded object. Since this was the first such dial to be recorded, Baldini could not have recourse to making comparisons. It should be noted, however, that the symmetrical marking out of the declination scales found on all other dials of this type is merely a stylistic feature. As can be seen in Figures 7b and 11, there is no need for a second scale on the opposite side of the centre, and there is thus no reason why the second

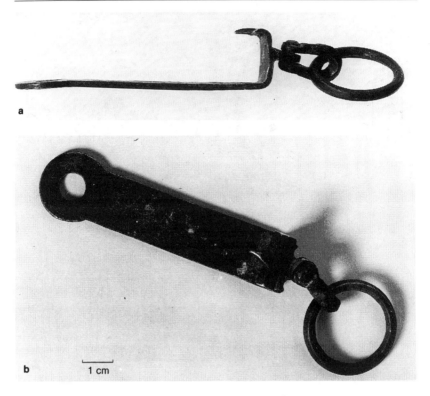

Figure 12 London sundial-calendar, suspension arm. Collection Science Museum, London, Inv 1983-1393. (a) Front view. (ScM neg. no 954/83/12); (b) side view. (ScM neg. no 1124/84/1) (*Reproduced by permission of the Trustees of the Science Museum*)

scale should not have been made slightly different from the first.

All three of the Latin dials agree in giving dates as an indication of how to set the gnomon/scale to allow for the Sun's declination (on the scales corresponding to $T_1 T_2$ in Figures 7b and 11).

The changing declination is a function of the Sun's changing position in the zodiac and expressing this in terms of the date is a realistic proposition only if one uses a suitable calendar. The surviving Latin dials all use the Julian calendar (introduced in 38 BC) but mark actual dates only for the two solstices: VIII K IAN (25 December) and VIII K IVL (24 June). Since the graduations on the date scale are very strongly non-linear, and the only date markings are those which divide the zodiac signs (which do not fall on the same date in each month) it seems that the user might need a certain amount of practice in learning to set the gnomon/scale correctly. Perhaps the purchaser of such a dial was given a small scroll of instructions for using it?

The dates these dials show for the solstices correspond to a spring equinox of 24 March, which is correct for about 150 BC. In fact, we know from

Ptolemy (*Almagest* III, 1) that Hipparchus observed the spring equinox for 146 BC to be 24 March (possibly 25 March by civil reckoning, since astronomical days began at noon). Nonetheless, 24 March seems to have been the date conventionally accepted for the equinox over a long period. It is, for example, quoted by Pliny as the date for the equinox in his own day and the same date is found on a fragment of an anaphoric clock for which Neugebauer accepts a date of *c.* AD 250.[15] Since the three Latin dials all use the Julian calendar they clearly cannot date from the period for which their equinox would be astronomically correct. It is, however, possible that the design of the instrument, which involves quite considerable understanding of astronomy,[16] did originate in that period and that the design was handed down merely as a design, being copied without modification. On the other hand, if the original designer merely marked the extreme lines as corresponding to the solstices, then conventional dates may have been introduced at any period. The use of conventional dates would be entirely comprehensible if the maker's customer were an educated layman, such as Pliny, rather than a professional astronomer.

THE DIALS WITH INSCRIPTIONS IN GREEK

The five dials with inscriptions in Greek (Figures 13–17) all resemble the Latin dial from Crêt-Chatelard in having a quadrant scale fixed with respect to the date/declination scales and running clockwise from zero to 90° in the

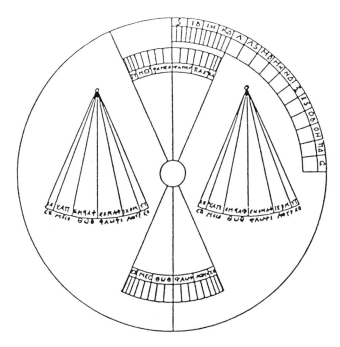

Figure 13 Early Byzantine portable sundial found at Memphis, reconstruction, after Buchner, 'Antike Reiseuhren' (note 5).

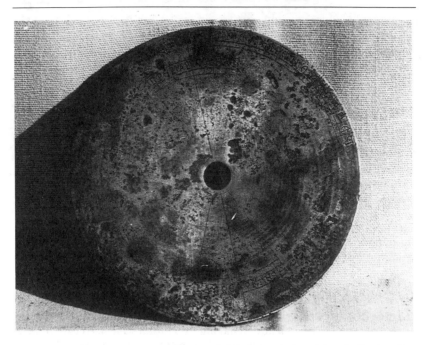

Figure 14 Early Byzantine portable sundial found at Aphrodisias. Collection Site Museum, Aphrodisias. (*Reproduced by permission of the Adler Planetarium, Chicago*)

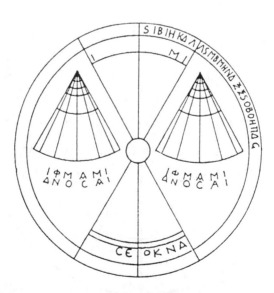

Figure 15 Early Byzantine portable sundial found on Samos, after Buchner, 'Antike Reiseuhren' (note 5).

quadrant starting from the line marking the equinox (Figures 2 and 13-16). The method of adjusting them for use at a particular latitude thus corresponds to what is shown in Figure 7c, and involves moving the point of suspension of the dial against the quadrant scale. As we have seen, for the Crêt-Chatelard dial Durand and La Noë postulated the use of a suspension arm (Figure 10). The only dial for which such an arm has been preserved is the London sundial-calendar. The arm is shown separately in Figure 12. We may note, however, that since the London instrument was considerably larger than the others, and much heavier (on account of the gearing), it is possible that the design of the arm may not be quite typical. In particular, its somewhat elaborate shackle and ring system contrasts with the simple eye supplied on the Oxford dial. However, a very similar shackle, though of less sturdy construction than that on the London dial, is found on the small ring dial discovered at Philippi (dated to the fourth century of the Christian era).[17]

Like the Latin dials, the Greek ones all mark their declination scales according to dates. The names of the months are given against the appropriate parts of the scale and on some of the dials some or all of the months are subdivided into thirds. The scales are thus more detailed than the corresponding scales on the Latin dials and we might perhaps suppose that this indicated a greater concern for precision. It is accordingly somewhat startling to find that the date for the equinox implied by the divisions on all but one of the five dials is 31 March. This would be astronomically correct for about 1450 BC. Readers with a taste for the bizarre may care to entertain the hypothesis that designing this type of dial was one of the achievements of Akhnaten's reform of Sun worship. As has been argued in detail elsewhere,[18] it is possible to avoid such a drastic reappraisal of New Kingdom astronomy. As already noted, only four of the five Greek dials give an equinox of 31 March. The remaining dial, the one found at Memphis (Figure 13), has its declination scale marked with the Egyptian months (presumably those of the fixed, Alexandrian, calendar). The equinox is the division between Phamenoth and Pharmouthi. I suggest that the strange equinox on the other Greek dials may have arisen from a crude 'Julianizing' of the scale, so that the division between Phamenoth and Pharmouthi (26 March) becomes the division between March and April (31 March). Connecting the Greek dials with the Egyptian calendar should not, however, be seen as implying that their design has any special association with Egypt itself. The Egyptian calendar, convenient because of the fixed length of its years, was in widespread use after as well as before the introduction of the Julian calendar. Nor does it seem that the equinox of 26 March found on the Memphis dial should be seen as marking any important difference between the Latin and Greek groups of dials. The subdivision of the months round the equinoxes on the Greek dials is into thirds (corresponding to the division of zodiac signs into decans). So if the equinox were supposed to be 24 March, the nearest division would be that between Phamenoth and Pharmouthi. Thus its adoption as the effective equinox may be considered a matter of simple rounding rather than a conscious use of approximation.

Astrological papyri show that, repeated decrees to the contrary not-

Figure 16 Early Byzantine portable sundial, sundial face. Collection Time Museum, Rockford (Ill.), Inv 1417. (*Reproduced by permission of the Time Museum*)

withstanding, the inhabitants of Egypt continued to use the movable (i.e. Ancient) Egyptian months as well as the fixed (Alexandrian) ones—with the same names—for several centuries after the official adoption of the Julian calendar.[19] If we are correct in our interpretation of the equinox date found on the Greek dials, then we have here further evidence that the Julian reform carried less weight, and was perhaps less well understood, in the Eastern part of the Empire than in the Western.

Dating the dials

None of the surviving dials of this type carries a maker's or owner's name or a date. They nonetheless all have quite extensive inscriptions in the form of lists of provinces and towns together with their geographical latitudes.

The lists of places on the Greek dials all include the name Constantinople, thus providing a definite *terminus post quem* of AD 328. Unfortunately, one cannot hope for such unambiguous evidence for a *terminus ante quem*. Since

Figure 17 Early Byzantine portable sundial, latitude list. Collection Time Museum, Rockford (Ill.), Inv 1417. (*Reproduced by permission of the Time Museum*)

these lists are not official documents they may well include names in their common rather than their official form, so that we cannot, for instance, argue that the use of the name Antiochia on the Aphrodisias, Rockford and London dials indicates that they were made before Justinian I (AD 527-565) refounded the city as Theoupolis in AD 542.

The lists on the Greek dials are substantially different from one another. Indeed, they give every indication of having been drawn up in accordance with the preferences of a particular customer. This is confirmed in the case of the Rockford dial by the fact that the latitude list on the back of the dial, shown in Figure 17, has been engraved much less neatly and much less deeply than the markings on the sundial face, shown in Figure 16, suggesting very strongly that the work was done by a different hand, or at least on a different occasion. (The style of the lettering is, however, closely similar, so there is no indication that the two sides were engraved at widely separated dates.) The list on the Samos dial is particularly idiosyncratic. As

Buchner has noted, it includes almost exclusively places in Asia Minor, showing that portable dials were made for people who had no apparent intention of travelling very far.[20]

Indeed, the list on the Samos dial even omits Rome, though the name is found on all the other Greek dials. It is thus clear that the latitude lists on such dials were not intended to give a representative list of important places throughout the Empire. In fact, it does not seem possible even to assume, for the purposes of dating, that all the places named on any of the Greek dials must have lain within Imperial territory at the time that dial was made. For even if we set aside Rome, as being of such cultural significance that its inclusion might be regarded as (almost) automatic, we find ourselves in difficulties concerning the interpretation of the latitude list on the dial found at Aphrodisias.[21] The list includes Merida (in Spain) and Burdigala (i.e. Bordeaux), places not included in the Byzantine Empire even after its drastic enlargement by Justinian's conquests. The inclusion of Burdigala gives the more stringent dating: a *terminus ante quem* of AD 349. This is about one and a half centuries earlier than the date indicated by the slightly spiky script with its scattering of serifs on 'C's, 'A's, etc.[22]

A fourth-century date for the Aphrodisias dial would also be at variance with the date suggested by a rather cruder indicator, namely the latitude given for Constantinople. Ptolemy's *Geographia* gives this latitude as 43°, but the value was changed to 41° by later geographers (the modern value is 41°02'). We find 43° on the Memphis dial, which Tischendorf dated (without explanation) to the fourth century. The same value appears on the Samos dial and on the dial now in Rockford (whose latitude list is given in full in Table 3). The remaining two dials give the latitude of Constantinople as 41°. This division into an earlier group (Memphis, Samos, Rockford) and a later one (Aphrodisias, London) is confirmed by the styles of the script on the various dials.

Subsidiary dials

Such a division is also supported by the fact that the three apparently earlier examples not only have the sundial scales of the type already described but also include a pair of subsidiary dials (Figures 13, 15 and 16).

On the Rockford disc there appears to be no material difference between the two subsidiary dials. Each shows the same pattern of lines, and measurement of their positions gave substantially identical results. The positions of the hour lines on these dials confirm Buchner's interpretation of the corresponding dials on the Samos disc (and his interpretation of Tischendorf's faulty drawing of the Memphis disc), namely that these subsidiary dials are in effect flat versions of the gnomon/scale used on the main dial.[23] If the suspension point of the disc is adjusted for the observer's geographical latitude then the radial lines on the two subsidiary dials indicate the directions in which the shadows of their gnomons are to fall for particular dates (the dial being turned appropriately). The lines themselves indicate the divisions between the months and thus correspond to the radial lines on the declination scales of the central dial (also in each case marked in months). In fact, on the Rockford dial one can make out traces of further, slightly lighter, lines dividing some of the months into three parts, as the months are divided on the declination scales of the central dial.

The flat hour scales on these subsidiary dials are even more severely non-linear than the curved hour scale on the Oxford dial. However, as we have noted above, there is no reason to suppose that a contemporary user would have been troubled by this non-linearity. As Buchner has suggested, introducing a curved form for the hour scale has the simple advantage of allowing the inclusion of a marking for noon, so that such a change may be regarded as an improvement, and the flat scales may be supposed to have been the earlier form of the dial.[24] The inclusion of a pair of flat dials—which seemingly cannot have been used in any way to check the setting of the main dial—may thus be an example of the conservatism found in the design of many later instruments. Redundant multiplication (here presumably for symmetry) is of course also a feature of many later instruments (such as compendia), and is particularly characteristic of instruments that are scientific merely in the sense of incorporating scientific results rather than in the sense of being usable to further scientific research.[25]

Evolution of the dials

As already noted, all the Greek dials give the latitude of Constantinople. None of the Latin dials does. In fact, the Latin dials list provinces rather than towns, and only the Oxford dial includes the name Rome. Names of provinces are slightly more difficult to interpret as evidence of date than are names of towns. Nonetheless, the fact that all the Latin dials mention Britain, Spain and Gaul as well as Eastern provinces such as Egypt suggests quite strongly that this group of instruments dates from before the collapse of the Western Empire and probably from before the division of the Empire in the fourth century. The latitude lists on the three Latin dials are all closely similar, though the Oxford instrument has some additional names not found on the other two.[26] It thus seems rather unlikely that even a detailed analysis of the lists would enable one to put the three dials into chronological order with any degree of confidence. Nor, *pace* Price,[27] does it seem very safe to suppose that the allegedly neater form of adjustment by means of a movable suspension arm, which was probably a feature of the Crêt-Chatelard dial, necessarily represents the more evolved form. It is, indeed, the form of adjustment adopted on all the Greek dials, whose date seems to be later than that of the Latin group, but the design of the Oxford dial (and probably of the dial found near Rome) in which one disc is recessed to receive another, is in fact reminiscent of the design of another later instrument, namely the planispheric astrolabe.

The Greek dials all postdate the division of the Empire (since their latitude lists include the name Constantinople) and therefore seem to be later than the Latin dials. As we have already noted, Tischendorf dated the Memphis dial to the fourth century. We are indebted to Professor Cyril Mango (Oxford) for suggesting, from an inspection of photographs, that the styles of their script place the Rockford dial in the fifth century, the London one in the late fifth or early sixth,[28] and the Aphrodisias one in the sixth century. The Samos dial has merely been dated to the early Byzantine period[29] but its latitude of 43° for Constantinople, and perhaps also its inaccurate value of the obliquity of the ecliptic, namely 30°, suggest a relatively early date.

These datings do not show any clear pattern of evolution for this type of dial. For instance, if we wish to accept Buchner's otherwise attractive suggestion that the subsidiary dials found on the three earliest Greek examples represent the original form, we must explain why this original form appears to have persisted only in the Greek dials and is not found in the Latin examples, which seem to be earlier, and one of which is of exactly the same design as the two much later Greek examples. Was the conservative force of tradition stronger in the Eastern Empire? It seems rather unlikely that we have separate Latin and Greek traditions. Moreover, as Buchner has shown, Vitruvius seems to regard the dials as Greek in origin, since he gives them the Greek name πρὸς πᾶν κλίμα.[30] If the design orginated a little before Vitruvius' day then it seems likely that the dials did indeed originally employ the Egyptian calendar, and we shall thus be able to explain away the absurd equinoxes found on four of the five Greek dials (see above). So far so good. However, it is a little awkward to claim the design as probably Greek in origin when the earliest surviving examples are all inscribed in Latin, and the original form is to be found only in subsidiary dials on later instruments whose main dials are adjusted slightly differently from two out of three of the Latin examples. Perhaps further archaeological discoveries will bring clarification.

WHY A SUNDIAL-CALENDAR?

Apart from the five portable sundials we have discussed, the only other Byzantine 'scientific' instruments known are a ring sundial datable to the fourth century,[31] a pair of dividers and an eleventh-century planispheric astrolabe[32]. The archaeological evidence, such as it is, thus suggests that the type of sundial we have been considering was at least a fairly common instrument.

The relative ordinariness of the London sundial-calendar is confirmed by our failure to detect any sign that our four fragments were gilded or silvered. Our investigation technique was by means of X-ray fluorescence using a microprobe in a scanning electron microscope and we are thus fairly confident that we should have found traces of such heavy metals as gold or silver had they been present. In fact X-ray fluorescence without the aid of an electron microscope served to establish that our Moon disc was tinned. Moreover, the teeth of the gears are not cut with great neatness. As can be seen in Figure 18, the filing sometimes departs from the scribed lines that show the division of the Moon disc into fifty-nine, and some of the teeth are canted over. Nevertheless, there is considerable evidence of wear, so we have every reason to suppose that the mechanism did actually work. This in turn suggests that the workman who made it had the measure of his task, that he was, in fact, engaged upon what he regarded as quite ordinary business. Indeed, if we think of the Antikythera mechanism—the only other material evidence we have for a Hellenistic tradition of mathematical gearing—the London calendar is by comparison extremely simple (even if

Figure 18 London sundial-calendar, Moon disc. Collection Science Museum, London, Inv 1983-1393. (ScM neg. no 954/83/4) (*Reproduced by permission of the Trustees of the Science Museum*)

one opts for the most elaborate of Wright's reconstructions). He and I are still of the opinion we expressed in a previous paper, that gearing was fairly commonplace and that as more Byzantine sites are dug more Byzantine gearing will emerge.[33]

As already mentioned, the outputs of the gearing of the London instrument certainly included the age of the Moon, in days, and an impression of its shape, as well as, very probably, the positions of the Sun and the Moon in the zodiac. The connection of any of these with the sundial part of the instrument is not obvious. An astronomer (or a graduate of the university of Alexandria?) might, indeed, seize upon the position of the Sun in the zodiac as giving a better way of setting the gnomon/hour scale against the declination scale than the crude scale of the Julian months actually marked on the front of the sundial. (The divisions marked on the declination scale do in fact correspond to the zodiac signs, some subdivided into decans—and we may note in passing that the division of some of the months into thirds is at variance with the use of the week in the gearing.) As Wright and I have suggested elsewhere,[34] it is possible to deal in the same way with the Moon, using its position in the zodiac to set the gnomon against the declination scale, so that the sundial can be used to tell the time by the Moon. Such a use would have involved some further approximations, and we have no way of knowing whether they would have been acceptable. For instance, obvious 'errors' in the readings of the dial might well be ascribed to the user rather than to the design of the instrument.

The ratchet, however, suggests that such elaborate uses for the outputs may be inappropriate. Since it prevents the gearing from being turned backwards, the ratchet makes it a little more awkward to use the instrument for computation. One would either require a method of releasing the spring against which the ratchet worked (so that the mechanism could turn backwards after all) or it would be necessary to take the instrument apart after each computation in order to reset the dials. In fact, the incorporation of a ratchet suggests that the mechanism was intended to be used by someone who preferred other people to do his thinking for him to the largest possible extent. It is perhaps worthy of note that this piece of idiot-proofing is not a feature of the apparently otherwise very similar mechanism which al-Bīrūnī describes as 'The Box of the Moon'.[35]

Despite the awkwardness introduced by the ratchet, it is difficult to believe that such calendrical gearing was not used for computation. The device would, for instance, make it very easy to convert from the popular lunar calendar into the official Julian one. The gearing could simply be turned on to, say, the next full Moon and the equivalent Julian date could be read off from the dial showing the position of the Sun in the zodiac. The other straightforward use of the gearing is that implied by the ratchet: the mechanism was turned forward each day—the click of the spring indicating that it had been turned sufficiently far—and the various outputs then served as a calendar.

In attempting to explain why a calendar should have been attached to the sundial we are perhaps creating a problem where none truly exists. For many sundials can in effect be used as calendars. Such a use is, for instance, obvious in the case of the subsidiary dials found on the Memphis, Samos and Rockford dials (see above, Figures 13, 15 and 16). Moreover, on a quite different scale, Buchner's description of the huge sundial set out by Augustus shows that its significance as a monument depends at least partly upon its being regarded as a calendar, the shadow of the tip of the gnomon moving along significant lines on significant days.[36] It thus seems possible that the maker of the London instrument merely decided to emphasize the possible calendrical uses recognized as already inherent in the sundial, and did so by adding a mechanical calendar. In seeking an alternative explanation for the combining of the two devices, we must ask ourselves why sundials themselves were made.

To answer such a question requires a grasp of the nature of Byzantine society, to which I can make no claim. Byzantinists have, however, suggested to me that portable sundials may have had a military application.

Manuals of military strategy frequently recommend that an army be divided into two parts which then make simultaneous attacks from different quarters. In such circumstances, the usefulness of a pair of matching sundials is obvious. The London sundial-calendar might then be the commander's dial, with gearing that would enable him to decide, for example, whether that night's Moon would be large enough to allow scouts to spy out the terrain. It was further suggested that such instruments might have been made by armourers, who were accustomed to doing fairly delicate work in base metal.[37]

FROM SUN TO STARS

Sundials of the type we have been discussing have, it seems, no place among Islamic astronomical instruments. However, the geared calendar associated with the London dial does appear: in the form of al-Bīrūnī's description of what must have been closely similar gearing (written about AD 1000), and as rather different gear trains, but with similar purpose, attached to an otherwise conventional Persian astrolabe dated AH 618 (= AD 1221/2) now in the collection of the Museum of the History of Science, Oxford (Computerised Checklist of Astrolabes, no. 5). Neither of these calendars is designed so as to have any necessary connection with an astrolabe. Al-Biruni's device is an entirely independent instrument (though the treatise about it is included in a collection that is very largely concerned with astrolabes[38]). The Oxford instrument consists of two devices back to back, housed on opposite sides of the central panel of what is in effect a mater with an I-shaped cross-section.[39]

This association of the calendrical gearing with the astrolabe is easily explained by the fact that the gearing has astronomical outputs and the astrolabe is the commonest astronomical instrument of the time. There are, however, more specific connections. Since the Islamic calendar is lunar, the position of the Sun in the zodiac is not so simply related to the date as it is in the Julian calendar. (As remarked above, it seems possible that the design for the Byzantine geared calendar originated from an instrument that employed, say, the Egyptian calendar.)

The position of the Sun is required in order to use an astrolabe to tell the time, which is one of the first uses described in treatises on the astrolabe and was very probably one of its principal uses in practice. The position of the Moon is also obviously useful, for instance for finding how many hours of moonlight remain. Furthermore, the positions of both luminaries play important parts in astrological calculations.

It is therefore not difficult to see why gearing like that of our Byzantine sundial-calendar should have proved to be of interest to Islamic astronomers. Why did they not also adopt the sundial? By this question I do not mean to imply that they could not have come across the gearing independently of the type of sundial to which it is attached in the so far unique surviving example. My point is that we have here rather hard evidence that Greek ingenious astronomical devices were known to Islamic scholars.

Among Islamic instruments, the most likely candidate for taking over the function of portable sundials is probably the planispheric astrolabe. The astrolabe is a very adaptable instrument compared with our sundial. However, although it can be used to perform a wide variety of different tasks, it does lack one kind of adaptability which is a prominent feature of this particular type of portable sundial. The astrolabe can only be used at latitudes for which that individual instrument has been supplied with plates. The sundial can be used at any latitude. (In both cases the observer must, of course, know his latitude—hence the tables which appear on both types of instrument.) Now, it seems possible that this particular form of adaptability may have seemed important enough to ensure the survival of such sundials at least for some time after the introduction of the planispheric astrolabe. So the existence of such dials in the fifth and sixth centuries should probably not be seen as evidence that the planispheric astrolabe was not yet in use. In fact, when complete the London sundial-calendar must have looked a little like an astrolabe—suspended vertically and with a gnomon to be adjusted rather than an alidade. In attempting to consider its wider context we might do well to consider the characteristically bold suggestion put forward in conversation by the late Professor Derek Price: 'Well, it's one of a pair of instruments. Where's the astrolabe that goes with it?'[40]

APPENDIX: THE ROCKFORD DIAL

The Rockford dial survives merely as a circular dial plate. Both the gnomon/hour scale and the suspension apparatus are missing. The instrument has no precise provenance but was allegedly found together with four other metal objects, apparently of similar date, identified in the dealer's catalogue as an alidade, a circular plate (with graduated rim) and two triangular set squares (one merely a drawing instrument, the other apparently designed to stand upright).[41]

Photographs of the two faces of the dial plate are shown in Figures 16 and 17. The plate is not exacly circular, but has a mean diameter of about 121 mm. The thickness of the disc varies considerably on account of the degree of corrosion, and perhaps also on account of some original irregularity. The mean thickness is about 3.1 mm. The diameter of the central hole is about 8 mm.

The sundial face
The sundial face of the disc is extensively corroded, and light-green corrosion products have accumulated in many of the engraved lines, making them relatively easy to distinguish. Some corrosion products have been removed, leaving pitting, and there are signs of scuffing (in parallel lines) which suggests that the surface has been cleaned. The engraving of all the inscriptions is deep and even.

The rim scale, running clockwise from zero to 90°, is numbered in intervals of 5°, each interval being subdivided into single degrees. At the 43° marking there is a neat pit, which is probably a punch mark and may be intended to indicate the point on the scale corresponding to the latitude of Constantinople. The double fan scale across the centre of the dial is marked with abbreviated forms of the Greek versions of the names of the months of the Julian calendar. Among the twelve names there are three scribal errors:—a rho for an iota in the name for May, a pi for a tau in the name for October, and an omicron for an epsilon in the name for December. The lines of the calendrical scales are engraved less deeply than the inscriptions and in places have been lost. It seems, however, that all the months were originally subdivided into thirds. As on the London dial, the subdivisions do not seem to be very accurate, though the distance from equinox to solstice, measured against the rim scale, is correct, namely about 23½°, and the total angle of the fan, measured with a modern protractor, is very close to 47°.

The two subsidiary dials on either side of the main one seem to have originally been more or less identical, though some of their lines have now been lost. At the vertex of each subsidiary dial there is a square area in which the surface of the disc is very rough. The patch near the left dial takes the form of a rough pit, whereas that on the right is partly raised above the level of the main surface. The diagonal of each square is aligned with the central radial line of the corresponding dial, and the corner is in each case very close to the vertex of the fan (where there is a punch mark on the left dial). It seems likely that these squares mark the positions of the bases of gnomons. The axis of each gnomon would presumably have been a line perpendicular to the disc through the vertex of the fan, and we may note that in consequence the shape of these gnomons would have to have been markedly different from that of the gnomon/scale on the Oxford sundial (see Figures 5 and 8).

There is no reason to suppose that the Rockford dial was intended to be a precision instrument, and my measurements of the scales of the subsidiary dials were consequently undertaken solely with the aim of discovering whether the nature of the scales was in accordance with Buchner's suggestion that the subsidiary dials were flat counterparts to the curved hour scale on the (now missing) gnomon/scale used on the central dial.[42] In taking my measurements I made no attempt to achieve great accuracy and merely employed an ordinary ruler, marked in millimetres.

Although traces of all the lines corresponding to divisions between the months could be seen on both subsidiary dials, the outermost lines and the central ones were considerably clearer than the others, so it was lengths along these lines that I chose to measure. Tables 1 and 2 show the lengths in millimetres, measured along these lines from the vertex of the fan to the arcs marked A, B, Γ, Δ, E (corresponding to the first to the fifth hours) (see Figure 17). Blanks indicate that the lines could not be seen at that position. Since the arc defining the fourth hour (Δ) proved particularly elusive, I made additional measurements on a line it could be seen to cross, namely the line next inside the left edge of the right dial. These are shown in the final line of Table 2.

Figure 19 shows the geometry of the subsidiary dials according to Buchner's reconstruction. The point O is the vertex of the fan, G the tip of the gnomon. The segments OA, AB, BΓ and so on all subtend angles of 15° at G. Therefore the line OΓ subtends 45° at G, that is $\angle \Gamma G O = 45°$, so the triangle ΓOG is isosceles. Therefore the length of the gnomon, OG, is equal to the length OΓ. As can be seen in Tables 1 and 2, all my measurements agreed in giving OΓ = 13.

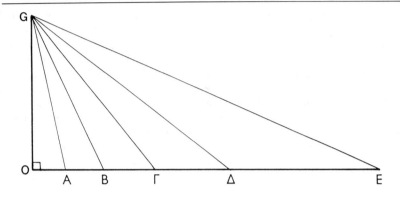

Figure 19 The arrangement of the gnomon and hour scale on the flat subsidiary dials, following Buchner, 'Antike Reiseuhren' (note 5). (*Reproduced by permission of the Trustees of the Science Museum*)

As can be seen in Figure 19, Buchner's reconstruction predicts that the distance of the line for the nth hour from the vertex of the fan will be given by the formula:

$$OH_n = OG \times \tan(n \times 15°)$$

Setting $OG = 13$, this formula gives the distances of the arcs marked A, B, Γ, Δ and E as 3.48, 7.51, 13.0, 22.5 and 48.5 mm (to three significant figures in each case). The agreement with my measurements is not perfect, but seems to me to be close enough to confirm Buchner's reconstruction, which was based on his study of the Memphis and Samos dials.

Table 1 The Rockford dial: measurements of the left subsidiary dial

Line used	Distance from vertex of fan (mm)				
	A	B	Γ	Δ	E
Left edge	3	7	13	–	48
Central line	3	7.5	13	–	48
Right edge	3	7.5	13	23	48

Table 2 The Rockford dial: measurements of the right subsidiary dial

Line used	Distance from vertex of fan (mm)				
	A	B	Γ	Δ	E
Left edge	4	7.5	13	–	48
Central line	4	7.5	13	23	47
Right edge	4	7.5	13	23.5	48
Line inside left edge	4	7.5	13	23	48

Table 3 List of latitudes on the Rockford dial, with the latitudes given for corresponding places in Ptolemy's *Geographia* and the modern values

				Latitudes	
				Ptolemy	Modern
ΜΕΡΟΗC	ΙϚ	Meroë	16	16°25'	[Khartoum 15°33']
COHNHC	ΚΔ	Syene	24	23°50'	24°05'
ΘΗΒΑΙΔΟC	ΚΗ	Thebaid	28		
[ΑΙΙΓΥΠΤΟΥ	ΛΑ	Egypt	31	[Alexandria 31°]	[Alexandria 31°]
ΠΕΝΤΑΠΟΛC	ΛΑ	Pentapolis	31		
ΑΦΡΙΚΗC	ΛΒ	Africa	32	[Carthage 32°40']	[Carthage 36°54']
ΠΑΛΑΙCΤΙΝ	ΛΒ	Palestine	32	[Caesarea Stratonis 32°32']	[Caesarea 32°30']
ΜΑΥΡΙΤΑΝΙΑ	ΛΔ	Mauretania	34		
ΚΥΠΡΟΥ	ΛΕ	Cyprus	35	[Paphos nova 35°10']	[Paphos 34°45']
CΙΚΕΛΙΑC	ΛΕ	Sicily	35	[South coast 36°]	
[ΡΟΔΟΥ]	ΛϚ	[?Rhodes]	36	[Rhodes 36°]	[Rhodes 36°26']
ΠΑΝ[Φ]ΥΛΙΑC	ΛΖ	Pamphylia	37		
ΕΛΛΑΙΔΟC	ΛΖ	Greece	37	[Athens 37°10']	[Athens 38°0']
CΠΑΝΙΑ]C	ΛΖ	Spain	37	[Malaga 37°30']	[Malaga 36°43']
ΤΑ[Ρ]COΥ	ΛΗ	Tarsus	38	36°50'	36°52'
ΑΝΤΙΟΧΙΑ	ΛΘ	Antioch	39	37°20'	36°12'
ΜΑΚΕΔΟΝΙΑ	Μ	Macedonia	40		
ΓΑΛΑΤΙΑC	Μ	Galatia	40		
ΘΕCCΑΛΟΝΙ	Μ	Thessalonika	40	49°20' [*sic*]	40°38'
Θ[Ρ]ΑΚΗC	ΜΑ	Thrace	41		
ΡωΜΗC	ΜΒ	Rome	42	41°40'	41°53'
ΙΤΑΛΙΑC	ΜΒ	Italy	42		
[Δ]ΑΛΜΑΤΙΑC	ΜΒ	Dalmatia	42		
ΓΑΛΛΙΑC	ΜΒ	Gaul	42		
ΚΑΠΠΑΔΟΚΙ	ΜΓ	Cappadocia	43		
Κ[ω]ΝCΤΑΝΤΙ	ΜΙΓ]	Constantinople	4[3]	43°05'	41°02'
[Α]ΡΜΕΝΙΑC	ΜΙΔ]	Armenia	4[4]		
Π[Α]ΝΝΟΝΙΑC	ΜΔ	Pannonia	44		
[Β]ΙΘΥΝΙΑC	[Μ]Δ	Bithynia	[4]4		
ΓΕΡΜΑΝΙΑC	[ΜΔ]	Germany	[44]		

The latitude list

The face of the disc that carries the latitude list is less well preserved than the sundial face. Its surface is irregularly and in places very heavily corroded, and is encrusted with red and green corrosion products. Despite the irregularity of the corrosion, the present surface is itself quite smooth, suggesting that some cleaning has taken place.

The depth of the engraving on this face is very uneven, ranging from lines as heavy as those of the inscription of the sundial face to lines that are

Table 4 **Comparison of latitudes.** The list of names from the Rockford dial is complete and in order. The Memphis dial has thirty-two names, the Samos dial twelve, the London dial sixteen and the Aphrodisias dial twenty-eight.[44] The names do not occur in the same order on the various dials.

Place	Latitude				
	Rockford dial	Memphis dial	Samos dial	London sundial-calendar	Aphrodisias dial
Meroë	16	16½	–	–	–
Syene	24	23½	–	24	23½
Thebaid	28	–	–	28	–
Egypt	31	[Alexandria 31]	–	[Alexandria 31]	[Alexandria 31]
Pentapolis	31	31	–	–	32
Africa	32	[Carthage 32⅔]	–	31	[Carthage 33]
Palestine	32	–	–	38 [*sic*]	36
Mauretania	34	–	–	–	–
Cyprus	35	–	–	–	34
Sicily	35	–	–	36	38 [*sic*]
[?Rhodes]	36	36	36	36	36
Pamphylia	37	36	–	–	36½
Greece	37	[Athens 37]	–	[Athens 36]	[Athens 37]
Spain	37	–	–	–	42
Tarsus	38	38	–	–	36½
Antioch	39	35½	–	36	35½
Macedonia	40	–	–	–	–
Galatia	40	–	–	–	42
Thessalonika	40	43	–	40	43
Thrace	41	41	–	–	–
Rome	42	41⅔	–	41	41½
Italy	42	–	–	–	–
Dalmatia	42	–	–	42	–
Gaul	42	44	–	–	46
Cappadocia	43	–	–	–	39½
Constantinople	4[3]	43	43	41	41
Armenia	4[4]	–	–	–	–
Pannonia	44	–	–	–	–
Bithynia	[4]4	–	–	–	–
Germany	[44]	–	–	–	–

apparently little more than scratches. The effect of unevenness has presumably been aided by corrosion, but in compensation colour contrasts in the corrosion products help to make certain parts of the inscriptions legible.

The face of the disc is divided by radial lines into thirty compartments, each of which carries the name of a town or province, written inwards from the rim. A circle concentric with the outer rim, and with a little over one-third of its radius, provides a margin for lining up the latitudes of the places concerned. Not all the names or latitudes are legible, but it appears that the places have been ordered according to increasing latitude, from Meroë at 16° to Germany at 44° or more. The place names, with their latitudes, are transcribed in Table 3, which also includes my identifications of the places concerned, and, where applicable, their latitudes as given in Ptolemy's *Geographia*[43] and in a modern atlas. Following the usual convention, doubtful letters are enclosed in brackets.

Table 4 shows a comparison of the latitudes given on the Rockford dial with those given for the same places on the four other dials of the same type with inscriptions in Greek.

Acknowledgements

I am grateful to the director and staff of the Time Museum, Rockford, Illinois, for making facilities available for a fairly detailed inspection of the Byzantine sundial in the Museum's collection.

References

1. J.V. Field and M.T. Wright, 'Gears from the Byzantines: a portable sundial with calendrical gearing', *Annals of Science*, Vol. 42, 1985, pp. 87-138. Reprinted in J.V. Field, M.T. Wright and D.R. Hill, *Byzantine and Arabic Mathematical Gearing*, Science Museum, London, 1985.

2. D.R. Hill, 'Al-Biruni's mechanical calendar', *Annals of Science*, Vol. 42, 1985, pp. 139-161. Reprinted in Field, Wright and Hill, op. cit. in Ref. 1 above.

3. See Field and Wright, op. cit. and J.V. Field and M.T. Wright, *Early Gearing*, Science Museum, London, 1985.

4. See D.J. de S. Price, 'Portable sundials in antiquity including an account of a new example from Aphrodisias', *Centaurus*, Vol. 14, 1969, pp. 242-266. The other 'scientific instruments' concerned are an astrolabe dated AD 1062, a pair of dividers (both mentioned in Price, op. cit.), and a ring sundial found at Philippi and apparently dating from the fourth century. See G. Gounaris, 'Anneau astronomique solaire portative [*sic*] antique découvert à Philippes', *Annali dell' Istituto e Museo di Storia della Scienza di Firenze*, Vol. V(2), 1980, pp. 3-18. This article originally appeared in *Ephemeris Archaeologike*, 1978 (1980), 181 ff. (in Greek). For the present purposes, a scientific instrument seems to be defined as a portable object, probably made of metal, with scientific connections. Of the instruments listed, only the astrolabe has specialized research potential. We shall return to this point below. See also Reference 25 below.

5. E. Buchner, 'Antike Reiseuhren', *Chiron*, Vol. 1, 1971, pp. 457-482.

6. Price, op. cit.

7. R. Tölle, 'Eine spätantike Reiseuhr'. In: *Archäologischer Anzeiger*, 1969, pp. 309-317; and Buchner, op. cit.

8. L.F.C. Tischendorf, *Notitia editionis bibliorum siniatici . . .*, Leipzig, 1860.

9. G. Baldini, 'Sopra un'antica piastra di bronzo, che si suppone un'orlogio da sole'. In *Saggi di dissertazioni accademiche publicamente lette nella nobile Accademia Etrusca dell'antichissima città di Cortona*, Vol. 3, 1741 (Rome), pp. 184-194.

10. V. Durand and G. de La Noë, 'Cadran solaire portatif trouvé au Crêt-Chatelard commune de Saint-Marcel-de-Felines [sic] (Loire)'. In *Bulletin et Mémoires de la Société nationale des Antiquaires de France*, série 6, 7, 1898 (Paris), pp. 1-38.

11. At the time of writing, the Oxford dial was on display at the Science Museum, London, as part of the exhibition 'Early Gearing' put on to mark the acquisition of the Byzantine sundial-calendar. For the catalogue to this exhibition, see Reference 3 above.

12. F.A. Stebbins, 'A Roman sun-dial'. In *The Journal of the Royal Astronomical Society of Canada*, Vol. 52, 1958, pp. 250-254.

13. Horam non possum certam tibi dicere: facilius inter philosophos quam inter horologia conveniet. Seneca, *Apocolocyntosis* [lit. 'Purgation'], 2.2.

14. This familiar result in fact depends upon the assumption, acceptable in Ancient times as in our own, that the radius of the Earth is negligibly small compared with the distance between the Earth and the Sun.

15. See O. Neugebauer, *A History of Ancient Mathematical Astronomy*, Springer-Verlag, Berlin and New York, 1975, p. 870.

16. Stebbins, op. cit. (Ref. 12).

17. Gounaris, op. cit. (see Ref. 4).

18. Field and Wright, op. cit. (Ref. 1).

19. F.K. Ginzel, *Handbuch der mathematischen und technischen Chronologie*, 3 vols., Leipzig, 1906. For references to more recent work, see B.R. Goldstein and D. Pingree, 'More horoscopes from the Cairo Geniza'. In *Proceedings of the American Philosophical Society*, Vol. 125(3), 1981, pp. 155-189.

20. See Buchner, op. cit. (Ref. 5).

21. Published in Price, op. cit. (Ref. 4).

22. I am grateful to Professor Cyril Mango (Oxford) for the information that the style of script on this dial indicates a date somewhere in the sixth century.

23. Buchner, op. cit. (Ref. 5).

24. Buchner, op. cit. (Ref. 5).

25. For example the polyhedral sundials preserved in many museum collections. A particularly elaborate dial of this type was made for the garden of the royal palace at Whitehall (London) in the seventeenth century. See J. De Graeve, 'L'Oeuvre gnomonique du P. Francis Hall (Linus)', *Bulletin de la Société Royale des Sciences de Liège*, 55e année, 1, 1986, pp. 207-219. See also J.V. Field, 'What is scientific about a scientific instrument?', *Nuncius*, Anno III, fasc. 2, 1988, pp. 3-26.

26. A comparison of the three lists can be found in Price, op. cit. The layout of Price's tables gives the impression that these additional names on the Oxford dial may occur as a separate list. They do not, as can be seen in our Figure 5.

27. Price, op. cit. (Ref. 4).

28. A more detailed discussion of the dating of this instrument is given in Field and Wright, op. cit. (Ref. 7).

29. See Tölle, op. cit. and Buchner, op. cit. (Ref. 5).

30. Buchner, op. cit. (Ref. 5).

31. Gounaris, op. cit. (see Ref. 4).

32. For the last two, see Price, op. cit. Some possible additions to this short list are mentioned in the section on the Rockford dial.

33. Field and Wright, op. cit. (Ref. 1).

34. Field and Wright, op. cit. (Ref. 1).

35. Hill, op. cit. (Ref. 4).

36. E. Buchner, *Die Sonnenuhr des Augustus: Nachdruck aus RM 1976 und 1980 und Nachtrag über die Ausgrabung 1980/1981*, Philipp von Zabern, Mainz, 1982.

37. These very helpful comments were made on the occasion of a brief visit to Dumbarton Oaks (Harvard University), Washington, D.C. in January 1985. The military connections were suggested by Drs Robert Edwards and Jonathan Shepard, the latter adding that the military would probably need the ratchet. The involvement of armourers was suggested by Professor Cyril Mango and others.

38. Hill, op. cit. (Ref. 2).

39. See Field and Wright, op. cit. (Refs 1 and 3).

40. In justice to Professor Price, to whom Mr Wright and I are very grateful for his valuable and enthusiastic assistance in the early stages of our work on this instrument, it should be pointed out that this remark was made before we had reached the conclusion that the sundial-calendar must date from the early Byzantine period (AD 328-641).

41. A. Brieux, *Histoire des Sciences, Livres—Instruments—Autographes* [sale catalogue], Alain Brieux, Paris, November 1977, p. 110 ff., item 9480.

42. Buchner, op. cit. (Ref. 5).

43. Ptolemy, *Geographia*, trans. E.L. Stevenson, New York, 1932.

44. For the Memphis dial see Tischendorf, op. cit.; for the Samos dial see Tölle, op. cit. and Buchner, op. cit. (Ref. 5); for the London dial see Field and Wright, op. cit. (Refs 1 and 3); and for the Aphrodisias dial see Price, op. cit.

Modern Construction Technology in Low-income Housing Policy

The Case of Industrialized Building and the Manifold Links between Technology and Society in an Established Industry

R.T. McCUTCHEON

INTRODUCTION

The housing problem
Since the Second World War there has been considerable discussion throughout the world about the imperative need to meet the rapidly increasing demand for adequate housing for low-income groups. The literature on the subject reveals differences of opinion regarding the scale of the problem, the priority to be accorded to it in a hierarchy of needs and the methods of solution.

In general it would be fair to say that the situation has been regarded as extremely serious. Mario Piche, the chief of Building and Infrastructure Technology Section of the United Nations Centre for Human Settlements (Habitat), wrote that it was conservatively estimated that one-quarter of the world's population did not have adequate housing and that in the cities of the developing world '50 per cent of the inhabitants, on average, live in slums and squatter settlements'.[1]

Partly as a result of the complexity of 'housing' and the 'housing problem' and partly as a result of the different ideological perspectives from which these issues have been viewed, many different policies have been proposed to solve the low-income housing problem: *laissez-faire*; financial measures such as lowering the rate of interest and increasing the loan replacement period; the facilitation of home ownership; wider provision of state housing; co-operative housing; expropriation and redistribution; self-help housing and sites and services.[2]

Modern technology: a new solution to the housing problem

While housing policies based on one or other of the concepts listed above have received varying degrees of attention since the Second World War, a major additional policy response on both sides of the traditional ideological divide has been to turn to modern technology for a solution to the problem. Modern technology, usually accompanied by reference to science and scientific method, has featured in low-income housing policy and related discussions in three different ways. First, reference has been made to the potential of modern technology as a whole, often with further generalizations as to the need for the industrialization of the building industry as a whole. Secondly, considerable attention has been paid to a set of methods of dwelling production which was variously termed prefabricated, industrialized, or systems building. (Here stress was laid upon the need for the large-scale mass production of dwellings by factory-based methods.) Thirdly, housing policies have been pursued which stressed a particular type of building, namely high-rise flats.

In his background report to a World Bank survey of low-income housing, Grimes was of the opinion that until recently most emphasis had been placed upon the need to increase the quality of housing and 'policies for the provision of housing centred on construction costs, combinations of alternative inputs, and the level of standards and finishes'.[3] In such a context it was anticipated that by using new materials, new architectural types, new methods of organization and new techniques of construction, a greater quantity of higher-quality housing could be produced using less-skilled labour, at a faster rate and for a cheaper cost than by using traditional types, materials and methods of construction. The belief was widespread and, as we will see, transcended the major socio-political divides. In 1966, a UN survey of government housing problems and policies categorically stated that 'whatever divergences of opinion might exist between different countries as to the necessity and scope of government intervention in the building field, there is a consensus of view that the production of buildings must increasingly adopt industrialized methods.'[4]

At the outset it is important to stress the scale of benefits that were anticipated. In her survey of innovation in the British building industry, Professor Marion Bowley remarked 'since the beginning of this decade [the 1960s] much has been heard of "industrialization" as the panacea for all troubles'.[5] More recently, in their *Housing of Nations*, Burns and Grebler reiterated this overview specifically in relation to industrialized building: 'until a few years ago, industrialized building systems were regarded as the panacea that would drive down costs'.[6]

Policies based on a wider use of modern technologies have been pursued and implemented for over thirty years. In relation to housing policy the questions are to what extent have they fulfilled expectations and what are the implications for current policies?

From the point of view of the relations between technology and society several aspects are of interest. These have been explored in relation to the development of high-rise flats in the UK[7] and in Iran,[8] and could fruitfully be studied on an international basis. Here we will focus upon industrialized building although, as we will see, the looseness of its definition can lead to a general discussion of industrialization as a whole, and the particular appropriateness of certain (much vaunted) industrialized methods to the construction of high-rise flats means that in some cases industrialized building results in high-rise flats.

Focusing upon industrialized building the relationship between technology and society will be explored under five headings: theoretical concepts and accompanying rhetoric; initiation of policy; implementation; comparison of results and implementation with theoretical concepts and its rhetoric; modification of the theory over time (but not the rhetoric).

THEORY, RHETORIC AND LINKS TO SOCIETY

Definitions of prefabricated, industrialized and system building are many and varied.[9] This has led to its meaning all things to all men and consequently it is difficult to tie down during argument. In addition, over the years, there has been a subtle transformation in the meaning of the concept. However, concentration upon the niceties of the distinction between one definition and another, or the precise point of demarcation between prefabricated and industrialized building, obscures a more important matter: both prefabricated and industrialized building stemmed from essentially the same theoretical premises and aroused very similar expectations. The guiding inspiration was, quite simply, factory-based mass production, in particular that used in the assembly-line production of motor cars. The major expectation was a phenomenal increase in production and quality at a reduced cost.

In the 1920s when mass-produced cars were first appearing in France, the architect, Le Corbusier, had asked 'Why not mass produced houses?'[10] and he spent much of his time over the next twenty years attempting to articulate his answers to this question. Others, such as Walter Gropius of the Bauhaus in Germany[11] and members of the Modern Movement in the UK, were similarly involved in theory and practice.[12] In the USA, during the 1930s, Albert Farwell Bemis wrote a trilogy, *The Evolving House*, which was a detailed exposition of what was entailed by the application of principles of mass production to housing.[13] Bemis introduced his theme as follows:

> For more years than I like to contemplate it has seemed to me that the means of providing homes in modern America and elsewhere have been strangely out of date . . . A new conception of the structure of our modern house is needed, better adapted, not only to the social conditions of our day but also to modern means of production: factories, machinery, technology, research.[14]

During the Second World War in the UK, the Committee for the Industrial and Scientific Provision of Housing (CISPH), which numbered such luminaries as Ove Arup (later Sir) and Joan Robinson amongst its members, advanced nearly identical arguments. The terms of reference for their work began with the statement:

> Technical development indicates the possibility of the manufacture of parts of houses by modern production methods with subsequent erection on site by assembly line methods such as are employed in other industries.[15]

The expectations aroused by the concept of modern technology and mass production may perhaps be gathered from the above quotations; these included the incorporation of modern facilities developed from scientific advances. The CISPH contended that 'the luxuries of a few decades ago were the commonplaces of 1939'.[16] Similar examples of concern for quality and modern conveniences may be found in the writing of Le Corbusier, Gropius and the Modern Movement.[17] Most importantly, mass production and modern technology would result in tremendous increases in quantity of production.

In his discussion of the principles of mass production in *Encyclopaedia Britannica* (1953), Henry Ford made a distinct link between mass production and a large-scale continuous assembly:

> The term mass production is used to describe the modern method by which great quantities of a standardized commodity are manufactured. As commonly employed it is made to refer to the quantity produced, but its primary reference is to method. In several particulars the term is unsatisfactory. Mass production is not merely quantity production, for this may be had with none of the requisites of mass production. Nor is it merely machine production, which may also exist without any resemblance to mass production. Mass production is the focussing upon a manufacturing project of the principles of power, accuracy, economy, system continuity, speed and repetition. The normal result is a productive organization that delivers in continuous quantities a useful commodity of standard material, workmanship and design at minimum cost. If production is increased, costs can be reduced. If production is increased 500 per cent, costs may be cut by 50 per cent, this decrease in cost, with its accompanying decrease in selling price, will probably multiply by 10 the number of people who can conveniently buy the product.[18]

When Bemis elaborated his arguments he made reference to the great improvements which had been made in other industries: 'a broad comparison of hand methods of a century ago with modern methods shows a phenomenal increase in output per worker—often 20, 30 or 50 times as much, in some instances more than 100 times'.[19] The Committee averred that 'the "Aladdin's lamp" that has brought about this strange and almost miraculous change is the machine'.[20]

In the UK a Temporary Housing programme was carried out in 1946 based on this approach. While a significant number of houses were produced only half the target was achieved at double the cost[21] and this somewhat tarnished the image of prefabrication for a while. When it resurfaced as industrialized building, we find that the turn of phrase adopted by the advocates of modern technology and mass production was little changed from forty years earlier. Industrialized building became the subject of numerous conferences in the late 1950s and 1960s. For example, the concluding statement of the Economic Commission for Europe's 1964 Seminar on the Building Industry included the following summary:

> It was unanimously agreed that the industrialisation of the building industry is necessary to overcome the gap between the needs of society and the capacity of the building industry.[22]

The notion that industrialized building would bridge the gap was often stated and more frequently implied. A 1965 United Nations report on the Effect of Repetition on Building Operations referred to the ever-growing demand for construction work of all kinds: 'the production level must in some cases be raised, not by 10 or 20 per cent, but by 200 or 300 per cent, if an acceptable rate of improvement is to be achieved'.[23] Although we see that this is down from the twenty, thirty to one hundred *times*, we are still observing claims of two to three *times*.

As mentioned earlier, industrialized building became the panacea which would solve the housing crisis. The transformation of the building industry by these means would have the effect best summarized in the conclusion and recommendations of a 1969 UN report:

> Industrialization of building is a necessity if rising demands for all kinds of construction work are to be met without a corresponding increase in the need for skilled and unskilled labour. Industrialization of building should at the same time serve the purpose of *improving quality*, *lowering costs*, and *speeding up the production*, of buildings in the same way as has already been achieved in the case of other industrially made products.[24]

Here we see not only the theory and the ethos, but also the option foregone of increasing the skilled labour. At the risk of over-repetition, the theory and its benefits were succinctly summarized in the words of another 1966 report which spoke of 'a new era, when mass production . . . will offer realistic possibilities for a rapid improvement of the housing situation'.[25]

The scope of the theory was international. The fundamental principles were seen to be applicable to both industrialized and developing countries. In a 1969 UNIDO report on the construction industry in developing countries, it was stated:

> large scale application of prefabrication methods rather than on site construction may prove to be the key not only to a rapid increase in building output but to increased employment of labour without impairing the quality of the end product.[26]

The use of capital-intensive methods with labour-abundant economies has not been seen by these commentators as a serious obstacle to the encouragement of housing policies based on modern technology: as in industrialized countries, 'housing agencies and officials have displayed intense interest in the concept of industrialised housing'.[27]

The rhetoric carried through into the 1970s and 1980s. In 1977 the survey of the policies of Eastern Europe and the Soviet Union for the period 1976 to 1980 stated:

> Turning to the construction process itself, it is noteworthy in the Soviet Union explicit targets for *cost reduction* are being set for building enterprises . . . the success of these latter measures must, however, be ultimately dependent on improvements in the actual technology and material basis of construction. The aim in its most general form is the *industrialization of building*, including complex supply and maximally economical utilization of the most advanced materials.[28]

Again, at a symposium held in Baghdad in 1976:

> In order to achieve this goal (of an activity five times higher than the present) industrialization of housing and building must play an important role in supplementing the traditional sector which cannot be developed to cover the needs of housing.[29]

In relation to developing countries the theoretical principles remain intact in relation to the ultimate objective, as has the rhetoric (yet examination shows the theory to have been significantly modified to take account of conditions in developing countries). For example, the general drift of the International Association for Housing Science is not just that industrialized building methods have 'worked in the European countries',[30] but also that the 'large experience gathered by countries which started earlier with building with large panels should serve the third world to plan properly their progress',[31] and industrialized building methods are considered 'a short cut remedy to their problems . . . by applying the know how and experience gained in many developed countries'.[32]

It would be fair to say that the majority of official pronouncements were along the lines outlined above. However, the theory has not been without its critics, who have felt that the analogy to the mass production of motor cars was not appropriate to the production of dwellings, either in process or product. In fact, reading the myriad reports on industrialized building one sees that many of its advocates were aware of possible shortcomings[33] and the traditional industry was increasingly adopting selected industrialized methods and thus becoming more efficient and for this reason offsetting the advantages of industrialized building.[34] An example of the awareness of the compromises which needed to be made in order for the current methods to be seen to accord with the principles may be illustrated by Professor Bowley's claim that

> the building site should be treated as an assembly line which differs from the assembly line in factories in that it is stationary, instead of moving and the workers, materials and other equipment, move round it.[35]

With respect to Professor Bowley, one wonders what Henry Ford would have thought about such a comment!

From the theoretical principles of prefabrication, some policies have aimed at the production of single-storeyed 'box' structures. Not only was such a theory advocated in the UK during the Second World War, but it was put into practice between 1945 and 1948. Another variant would be the mobile home, which formed a significant proportion of housing used by low-income groups in the USA during the 1970s. Equally, a limited development of mobile house production took place in the UK during the mid-1970s. But in both the USA and the UK, a mobile home fell outside the definition of a dwelling as laid down in the building regulations and the specifications adhered to by the building societies. Furthermore, this type of mobile home was *not* the focus of attention by policy makers and their advisers when discussing industrialized building.

In conclusion, although the definitions are varied and cover a wide range of building from rationalized-traditional (the application of scientific planning to traditional construction) through to industrialized-system building, the essential objective of these processes is that the housing should be produced as far as possible in a factory by methods which are in principle similar to those used in assembly-line mass production of commodities such as automobiles, washing machines and so on.[36] It was anticipated that the results would be an order of magnitude better than those achieved by traditional methods.

TECHNICAL ADVICE AND INITIATION OF POLICY

Policies based upon the same theoretical premises have been seriously advocated across the traditional ideological divide and in both the industrialized and developing worlds.

On a general level the literature revealed that there were various international forums which enabled and encouraged contact between prominent academics, practitioners and advisers to governments and international agencies. In 1959 the United Nations Economic Commission for Europe (UNECE) convened its first seminar devoted to the state of prefabrication in Europe; other seminars were convened during the 1960s by the ECE.[37] In 1964 the Housing, Building and Planning Committee of the UN's Economic and Social Council proposed the 'organization of Seminars and study tours on the development of industrialized buildings and related subjects'[38] as a standing item of the committee's programme of work. Since that time the committee has convened several international conferences which have been supplemented by roving seminars in Africa, the Middle and Far East, and Latin America.[39] For its part, UNIDO convened Expert Group Meetings on industrialization and prefabrication in the construction industry.[40] Since 1959 many conferences of this nature have been organized by the International Council for Building Research Studies and Documentation.[41] During the 1970s and 1980s the activities of these bodies have been supplemented by the Council on Tall Buildings and Urban Habitat[42] and the International Association of Housing Science.[43]

Over the years these international committees, councils and conferences have been attended by people holding senior positions in governments, the design professions, research establishments, international agencies and industry. They have represented countries across the social and political spectrum. Housing agencies from all the countries listed in Table 1, as well as others, have been represented. A distinct similarity of views was expressed. Through the published activities of these international organizations, it has been possible to trace many individuals who have remained committed to the propagation of industrialized building over a long period of time. These individuals might not consider themselves to form an identifiable group but their association with one another over a long period of time does make them a *de facto* core pressure group.

The existence of a theory is a necessary prerequisite for its propagation—and we can certainly see the channels through which this took place—but it is insufficient on its own to lead to either initiation or implementation. However, in this respect it is worth stressing the numbers of senior government housing ministers, housing advisers and directors of national institutes of building or architecture, who were involved. One section comprised policy makers themselves, while another consisted of the people to whom policy makers turn for advice regarding housing technology.

Four case studies, two based on previously published (though scattered) material and two on original research, revealed that while this was a dominant theory there were countervailing forces. Initiation of policy still required an entrepreneurial approach by committed individuals. In the Soviet Union, for example, the theory was fully established amongst the design professions by the time of the Second World War, but concentration upon prefabrication in general, and the 'progressive method' of large panel construction in particular, was due to intervention on the part of Premier N. Khrushchev.[44]

Table 1 Countries which have implemented programmes on principles of industrialized building or have seriously examined its potential

Australia	Italy
Belgium	Malaysia
Bulgaria	Netherland
Colombia	Norway
Czechoslovakia	Philippines
Denmark	Poland
Egypt	Puerto Rico
Fiji	Romania
Finland	Saudi Arabia
France	Singapore
German Democratic Republic	South Africa
Federal Republic of Germany	South Korea
Ghana	Sri Lanka
Hong Kong	Sweden
Hungary	Thailand
India	Union of Soviet Socialist Republics
Indonesia	United Kingdom
Iran	United States of America
Iraq	Yugoslavia

In the case of the UK it was possible to identify a section of the design professions who had long been committed to the production of dwellings by industrialized methods. These people gradually achieved positions of influence in central and local government and pressed for the adoption of industrialized building. In the case of the Ministry of Housing and Local Government, the technical advice accorded with the convictions of the progressively minded and powerful Permanent Secretary. Equally, during the early 1960s the Minister of Housing and the Minister of Public Building and Works had close connections with large building companies; while among certain large building companies consideration of the potential of industrialized building had increased.[45]

In the early 1960s, in the UK, there was thus an identity of interest amongst a powerful segment of those concerned with the production of public housing: large local authorities, central government, an influential segment of the design profession and certain large building contractors. The identity of interest was bound together by a common belief as to the nature of modern building, which was reinforced by contempt for traditional methods. The result was the industrialized building drive which was initiated in 1963 by the formation of various institutional groups such as the National Building Agency, the Directorate of Research in the Ministry of Public Building and Works and the Development Group in the Ministry of Housing and Local Government. The effect of these developments was accelerated in 1965 by the Ministry of Housing's circular which stated that by 1970, 40 per cent of all buildings had to be industrialized—and therefore a higher proportion in large local authorities. Such overt activity was supplemented by the personal efforts of senior officials in various ministries.

The provision of low-income housing in the UK is the result of action by various parties: government, local authorities and industry. While the user has been the ultimate reason for building the housing, he/she has very seldom taken an active part in the process of the provision of housing. Within a local authority, housing policy is decided upon by the Housing Committee. These decisions are usually taken after consultation with the various experts employed by local authorities who are responsible for the local authority side of the production process: the architect, planner, engineer, surveyor and valuer. In order to obtain a government loan the proposed housing has to be approved by the Department of the Environment. Once approval has been granted the project may be built by direct labour or it is put out to tender for construction by private companies. The involvement of these several parties means that up to a point there must be a certain amount of consensus for any new policy to be initiated. Therefore, it was difficult to identify one factor which was of overwhelming importance in leading to the introduction of the industrialized housing drive. However, while the following all played their part, the first was the most important during the period leading up to the initiation of the policy:

1. The theory of industrialized building was espoused by wide sections of the design professions; a particular group of members of the design professions, particularly architects, who were convinced of the potential of industrialized building, rose to positions of authority.

2. There was a perceived need for an increase in the production of housing without an increase in the traditional resource base, combined with an assessment that industrialized building would provide an increase in volume without requiring an increase in traditional skills.
3. Certain large building contractors anticipated that industrialized building would enlarge their building programmes and result in increased turnover and profits.
4. The development of high-rise building led to familiarity with a form of structure for which there existed the widest range of industrialized building systems; it was seen to be the most suitable form of structure for construction using industrialized methods.
5. Specific links between the government and the building industry encouraged the introduction of industrialized building.

These factors may be seen to apply to specific parties. A further factor served to bind together the actions of the various parties so that they achieved a consensus and developed a common policy: a widespread belief in the need to use modern technology in order to solve the housing problem. One of the most frequently used words in connection with industrialized building was 'inevitable'[46] and an oft-repeated phrase was the 'impossibility of putting the clock back'.[47] In 1963, at a conference on industrialized building, one leading contractor was reported as saying that ' "industrialized building" meant building in the manner of other things done in the 20th Century'.[48] A corollary of this thinking, which helped to obscure consideration of alternatives, was an antipathy towards the traditional building industry and an out-and-out rejection of its backward methods of production. For example in 1966, George Lothian, the General Secretary of the Amalgamated Union of Building Trade Workers, stated that 'too many firms are still in the "wheelbarrow-and-spade age" '.[49]

In the USA the theoretical premises of Operation Breakthrough, which was initiated in 1969, were identical to those expressed in Eastern Europe and the welfare economies of Western Europe;[50] the same was true of Iran's Crash Housing Programme from 1975 to 1978.[51] In both cases the initiation of policy stemmed from strenuous efforts by technical advisers.

IMPLEMENTATION: QUANTITATIVE ANALYSIS, EFFECT OF SOCIAL FACTORS, AND COMPARISON BETWEEN RHETORIC AND REALITY

Quantitative analysis: the broad effect of implementation within industrialized socialist, welfare and capitalist/market economies*

Policies based on the principles of prefabrication and industrialized building have been adopted by many different countries as varied as the USA (Operation Breakthrough), the USSR, Iran and Fiji (Table 1). However, the bulk of dwelling construction based on these policies has taken place in Western and Eastern Europe (including the USSR). Although it might seem obvious to those well acquainted with the field, it is considered useful to demonstrate using statistical data that usage of industrialized methods has

---

* This sub-section is taken from R. McCutcheon, 'Industrialised house-building 1956–1976', *Habitat International*, Vol. 12 (1), 1988, pp. 95–104.

Table 2 Percentage of public residential dwellings constructed by prefabricated methods (Western Europe)

Country	Type	1956	57	58	59	60	61	62	63	64	65	66	67	68	69	70	71	72	73	74	75	76
Austria	All types (of prefab)															10.7	10.7	10.7	10.9	10.8	10.7	9.4
	Multi-dwelling															16.0	18.5	18.9	12.8	17.7	17.2	15.2
Belgium	All types								1.8[1]													
Denmark	All types							21.4[3]				33[2]						29.5[4]				
	Large panel											35.7[a]						20.0[5]				
Finland	All types																					
	Collective							10[8]														
	Apartments											25.9										
	Large panel											15.7[a]		15[7b]								
France	All types							8.6	10.6	12.3	15.1											
	Large panel (ECE)			4–5[6]				5.5	6.6	7.6	7.1	17[7a]										
	Multi-dwelling:																					
	All types							11.2	14.3	14.9	18.5			25[7b]								
	Large panel							7.0	8.0	9.1	8.4											
FRG	All types										3.7[c]	4.7[c]			5.4	5.3	5.1	8.0	7.3		9.4	9.5
	Multi-dwelling															7.6	6.1	8.0	5.7		8.4	7.2
Italy	All types[10a]											1.2	0.5	0.5	0.4	0.4	0.5	0.5	0.5	0.5	0.6	
Netherlands	With aid of prefabrication[10b]		7.0	11.0	11.0	13.0	11.0	10.0	10.0	13.0	15.0	17.0										
	Large panel							2.0[12]	4.0[12]					9.7[7b]								
Norway	Share of prefab																	7	6.3		5.9	
	Light-weight concrete[11]																					
	All types			11.0	10.0	9.0		7.2	9.1	4.3	3.6	2.8	1.6	1.6	1.0	0.8	0.7	0.4	0.4	0.5		
	Multi-storey dwelling			25	23.9	18.9		13.0	23.0	4.7	6.0	1.8	1.1	1.2	0.7	0.3	0.5	0.3	0.5	0.3		
Sweden	Multi-dwelling all types prefab[14]			7.4[13]			4.2[13]	3.2[13]				12[15]										
	Large panel							2.6[3]				12[15]		14[7b]				18.5	18.7			6.3
UK[16]	All types											26.0	29.6	33.9	36.8	37.8	34.0	23.4	19.8	21.7	20.2	18.8
	Large panel											8.4	9.5	12.3	14.5	16.7	15.1	10.8	5.3	4.4	1.2	1.4
	Multi-dwelling																					
	All types											27.6	35.9	32.3	30.0	41.5	37.2	24.9	20.2	22.6	16.9	14.2
	Large panel														19.6	32.5	24.9	13.2	8.8	7.5	2.2	3.2

In this and the following tables, please note the cautionary statement at the start of the 'Implementation' section. Notes to this table are on p. 175.

Table 3 Percentage of public residential dwellings constructed by prefabricated methods (Eastern Europe)

Country	Type	1956	57	58	59	60	61	62	63	64	65	66	67	68	69	70	71	72	73	74	75	76	77
Bulgaria	All types prefab						2.7	4.0	4.0	7.3	11.9	16.7	18.7	16.9	25.6	23.4	31.7	31.9	38.5	52.9	59.5		
	Medium panel						1.2[1]	2.3[1]	2.9	6.7	10.2	14.7	18.2	14.1	23.2	22.0	29.1	27.4	30.4	33.4	39.3		
Czecho-slovakia	All types prefab					50[15]							63.4	66.9	68.5	68.0	67.4	67.2	68.3	70.1	70.3	66.9	
	Large panel												43.9	49.5	48.0	53.6	56.0	58.5	58.6	62.8	63.0	61.3	
	Multi-dwelling																						
	All types			15.3	19.2	31.7	46.4	54.7	59.9	75.0	77.4	78.5	84.6	86.7	91.2	92.3	92.5	92.9	93.5	95.8	95.8	95.4	
	Large panel			7.6[2]	9.6[2]	17.0[2]	26.0[2]	31.6	36.6	53.2	55.1	58.1	58.5	64.2	65	72.6	77.2	80.9	80.0	85.9	85.8	87.4	
GDR	All types prefab[3]								76.2	87.3	94.5	96.4	97.2	96.6	98.2	90	88.1	86.6	84.3	79.6	80.1	82.6	
	Large panel								18.0	23.2	22.4	35.4	37.4	39.8	44.4	54.4	61.2	63.5	61.6	62.9	62.9	63.5	
Hungary	All types prefab											10.4[10]										35.7	
	Large panel																28.3	32.3	35.2	35.9	38.0	28.2	
	Multi-dwelling																						
	All types						18.2[4]	39.2[4]	43.8	48.8	48.8	50.6	52	55.3	64.8	72.4	74.2	74.0	76.1	72.4	73.3	75.6	
	Large panel	3.0[5]					0.3[4]	1.3[4]	1.3	2.7	6.1	13.2	16.7	22.9	36.5	49.7	55.1	59.5	62.4	59.9	63.2	67.1	
Poland	All types prefab		0.1[6]			11.5[5]		24.4[5]	28.0[5]		38.0[5]												
	Large panel					3.0[6]			11.0[6]			19.6[10]											
Romania	Large panel								13.0[7]	13.0[7]		11.3[10]											
USSR	All types prefab						9.2[9]	8.0[18]	20.2[9]	14.4[17]	30.0[9]	65.7[12]	32.2[9]		35.8[9]								
	Large panel				1.3[9]	3.5[19]						32[10]									50[13]	50[14]	
Ukrainian SSR	All types prefab			35[11]				80[11]															
	Large panel					3.2[11]	9.8[11]	17.3[11]															
Yugoslavia	Non-trad.								3.8	13.0	2.9	5.6	4.1	3.7	3.2	3.4	4.0	2.7	3.4	2.8	3.2	4.0	

Notes to this table are on pp. 175–6.

generally increased as the proportion of all dwellings provided by the public sector has increased. Furthermore, the greater the use of industrialized methods as a whole, the greater has been the use of sophisticated methods of industrialized systems building.

Data on Industrialized Building

The proportions of housing built by prefabricated methods in selected countries in Western and Eastern Europe may be seen in Tables 2 and 3. In 1968 a similar table formed part of a report to a sub-committee of the US Congress. A note to that table warned:

> Figures above that are from different sources are based on different, often overlapping categorizations of building processes. The yearly percentages for each country are therefore non-comparable, non-additive and in some instances conflicting.[52]

In that report data were provided up to 1966. Since that time data have become available for Europe in the *UN Annual Bulletin of Housing and Construction Statistics*. Nonetheless, the data are still not strictly comparable, and, unfortunately, the proportion of housing built by industrialized methods in France and the USSR is not recorded. Other sources were resorted to for such coverage, with the result that the cautionary note of 1968 must still be borne in mind. Despite such provisos several observations may be made.

The role of industrialized building (both large panel and prefabricated component systems) in the provision of public sector residential dwellings is greater in Eastern than in Western Europe, in some cases considerably greater. Furthermore, while the role of industrialized building has decreased in Western Europe, it has steadily increased in Eastern Europe.

It is particularly noteworthy that the greater the role of industrialized methods as a whole the greater the use of large panel systems. In Western Europe, large panel construction accounts for a relatively smaller proportion of all the housing which has been classified as industrialized. Here again, the use of large panel systems increased throughout the 1970s in Eastern Europe whereas it decreased significantly in Western Europe. The differences may be illustrated by contrasting the experience in the USSR with that in the UK.

In the USSR the use of large panel industrialized construction increased from 1.3% in 1959 through 35.8% in 1969[53] to 50% in 1976.[54] For the five-year period from 1976 to 1980 the planned estimate was '60 per cent of all housing construction in the country'.[55] By 1990 the USSR plans to build 75% of all housing units by this method.[56] The proportion of large panel prefabrication was already significantly higher in large cities and new towns than in other urban areas: for example, Moscow—83%; Bratsk—84%; Togliatti—97%.[57] In addition to large panel construction methods, the use of more sophisticated box systems and less advanced prefabricated components (large block) means that actual reliance upon prefabricated methods is higher than that indicated by the figures relating to large panel construction alone. For example, in the early 1960s when less than 10% of state and co-operative housing was built by large panel methods, it was reported that a significantly higher proportion of all dwellings was built using more elementary systems of prefabrication.[58] In this respect other data indicate that the use of prefabricated components throughout building has increased from 30 million in 1960 to 119 million in 1976. During the mid-1970s the Director of the Central Research and Design Institute for Dwellings stated that 'not less than half the industrialized housing construction of the world is carried out in the USSR'.[59]

While the numbers are impressive, it must be pointed out that even in the USSR it has taken much longer than expected to institute industrialized methods. In 1959 it was predicted that by 1964 large panel methods would account for the majority of new dwellings,[60] yet these methods only accounted for 50% of urban housing by the mid-1970s. Furthermore, the type of prefabrication which was thought (in 1959) to hold the most promise—multi-storey 'box' construction—still remains on an experimental basis.

The general experience in the USSR contrasts strongly with that in the UK. Over the period from 1965 through to 1977, 26.6% of the dwellings built in the public sector, under the auspices of local authorities and new towns, were constructed using industrialized methods.[61] The use of industrialized methods reached a peak of 42% in 1969 and fell to 5% by 1977. Furthermore, large panel and other sophisticated systems were only used to construct about 36% of the dwellings built by industrialized methods; the majority of these were approved and constructed during the mid to late 1960s. In fact, the systems which were most widely used overall were of the *in situ* type (37%), with timber-based systems in second place during the 1970s. These were the systems which conformed closely to the traditional image of a British house.

The differences between Eastern and Western Europe as regards the role of industrialized building are even greater when we take into account the proportion of all dwellings constructed by the public sector. Tables 4 and 5 show the proportion of dwellings completed by the type of investor for countries in Western and Eastern Europe and for the USA. Plotting the proportion of public sector housing constructed using industrialized methods against the proportion of all dwellings built by the public sector shows that the reliance on industrialized methods is roughly proportional to the role played by the public sector in the provision of housing.

In sum, during the 1960s and 1970s the large-scale use of industrialized methods of building over an extended period of time was dependent upon the prior existence of a significant public housing sector. Furthermore, the extent to which the more sophisticated systems were used was related to the scale of reliance upon industrialized methods of building as a whole.

The effect of other social factors on implementation

The scale of development in the USSR is too great to be ignored. However, the development has been nowhere near as smooth as has been portrayed by official pronouncements on the subject. We have seen that the time-scale was underestimated. Furthermore, the necessary concentration of the housing market and building industry was brought about by political

Table 4 Percentage of dwellings completed by type of investor (Western Europe and USA)

Country	1963			1968			1973		
	Public	Housing assoc. or co-op	Private	Public	Housing assoc. or co-op	Private	Public	Housing assoc. or co-op	Private
Austria	11.4	28.9	59.7	9.9	37.5	52.5	11.4	29.6	59
Belgium	0.3	1.0 (1964)	98.7 (1964)	0.4	1.8	97.8	1.0	1.3	97.7
Cyprus						93.2 (1967)			100
Denmark	2.3	33.6	63.7	1.8	36.5	61.7	0.6	23.3	76.1
Finland	13.2	57.1	29.7	20.1	48.3	31.6	16.1	61.6	22.3
France	32.8		66.1	36.7		63.2	24.3		75.7
Greece			100			100			100
Ireland	30.3	69.7		36	64	43.2		73.6	53.3
Netherlands	19.8	24.2	56.0	23.8	32.9		5.8	40.9	
Norway	3.8	31.3	64.9	5.5	30.4	64.1	6.5	27.1	64.4
Portugal	8.3		91.7	6 (1967)		94 (1967)	10.9	0.6	88.5
Spain	14.2		85.8	8.8		91.2	8.7		91.3
Sweden	35.1	24.8 (Co-op)	40.2	44.3	19.3 (Co-op)	36.4	39.6	10.0 (Co-op)	50.3
Switzerland	2.1	9.4 (Co-op)	97.1	2.1	5.6 (Co-op)	92.3	4.1	8.4 (Co-op)	87.5
UK	41.9		58.1	46.8		53.1	36.8		63.2
USA	2.0		98.0	2.6		97.4	0.9		99.1
FRG	2.5	24.4	73.1	2.5	22.5	76.0	2.0	16.7	81.3
Turkey	1.0	1.9 (Co-op)	97.1	2.1	5.6 (Co-op)	92.3	4.1	8.4 (Co-op)	87.5

Source: UNECE, *Annual Bulletin of Housing and Construction Statistics 1971*, Vol. XV, Table 5; UNECE, *Annual Bulletin [...] 1974*, Vol. XVIII, Table 5.

Table 5 Percentage of dwellings completed by type of investor (Eastern Europe)

Country	1963			1968			1973		
	Public	Co-op	Private	Public	Co-op	Private	Public	Co-op	Private
Bulgaria	15.9	–	84.1	28.0		72.0	41.6		58.4
Czechoslovakia	48.4	24.3	27.3[1]	24.2	53.7	22.2[1]	44.2	29.6	26.2[1]
GDR	35.7	54.3	10.0	63.8	26.6	9.6	59.3	25.7	15.0
Hungary	37.6		62.4	36.7		63.3	32.9		67.1
			(38.2 aided)			(44.4 aided)			(54.9 aided)
			(24.2 not)			(18.7 not)			(12.2 not)
Poland	52.2	16.1	31.7	24.9	48.4	26.7	30.6	44.9	24.5
Romania	34.6		65.4	49.6		57.2	40.9		59.1
USSR	64.1		35.9	71.8		28.2	79.4		20.6
			(18.0 E)[2]			(12.6 E)[2]			(9.9 E)[2]
			(17.9 C)			(15.6 C)			(10.7 C)
Yugoslavia	39.6		60.4	34.0		66.0	33.2		66.8

1. Mainly in rural areas.
2. E = employees; C = collective (housing provided or subsidized to an employee by a company, and housing provided through a private collective respectively).

Source: UNECE *Annual Bulletin of Housing and Construction Statistics 1971*, Vol. XV, 1972, Table 5; UNECE *Annual Bulletin [. . .] 1974*, Vol. XVIII, 1975, Table 5.

action.⁶² The difficulties experienced suggest caution is necessary in relation to the following: the planning process; the organization of construction; the time-scale within which the production process may be revolutionized; and the quality of housing produced. These difficulties have been experienced despite the presence of a relatively concentrated building industry and continued political pressure over a thirty-year period.

Once the political intervention had been made, industrialized building made steady progress. In terms of concrete technology and capital investment involved, developments were more advanced than those in Western Europe. But, despite the consistent commitment to industrialized building and continued research and development, the much-vaunted multi-storey box system of the 1950s and 1960s, which was expected to be the ultimate form of prefabrication, has remained largely experimental. Thus, even under the most auspicious circumstances the developments have not run according to theoretical projections.

In the USA, Operation Breakthrough was initiated in 1969. It was intended that it would result in the wider use of factory-based industrialized methods along the lines developed in Europe. The main difference in intellectual terms was that whereas European programmes had been based upon the orthodox construction industry, in Operation Breakthrough it was hoped to interest the major manufacturing companies. However, the actual programme of work undertaken by the Department of Housing and Urban Development was very small—some 2,795 dwellings—and no further dwellings have been produced as a direct result of Operation Breakthrough. Although a parallel programme by the Department of Defense resulted in 3,160 dwellings, Operation Breakthrough did not achieve its primary aim of 1,000 dwellings per year over five years. In short, the policy maker's confidence in the theoretical bases for intervention by major manufacturing corporations, was not matched by the corporations themselves.⁶³

To this the retort may be that other political developments led to a massive increase of production within the traditional industry and the mobile housing industry, which alleviated the need and removed the incentive for intervention. However, reference to the growth in the mobile home industry indicates the discrepancy between a modern low-income dwelling technology indigenous to the USA, and its European equivalent.

The fate of Operation Breakthrough in the USA would reinforce the contention that industrialized production of housing—along European lines—is not possible unless there is a significant public housing programme. The nature of the technology involved requires government intervention in order to institute the necessary degree of co-ordination.

In Iran, the Crash Housing Programme initiated in 1975 failed to produce many of the proposed 200,000 dwellings over its 2½-year period. The very time-scale indicated the extent to which policy makers thought (or pretended) that a miracle could be achieved. More importantly the failure once again indicates the pre-eminent need for a public housing sector (in relation to industrialized building). The programme was based upon exhortation of the private sector and peripheral activity such as small demonstration projects. However, the established private sector construction industry was not at all involved in the provision of low-income housing.[64] More importantly, systems building was alien to the process whereby low-income accommodation is created in Iran—locally made, sun-dried mud-brick buildings. Moreover, the result of using the methods in 'large scale apartment complexes' would be even more unfortunate in the case of Iran than in the UK.[65]

However, it is clear from the Iranian case that despite its shortcomings and the social context necessary for success, industrialized building can still evoke policies aimed at a radical improvement in housing conditions in countries which are neither industrialized nor have large public housing programmes.

In the UK the industrialized building drive of the 1960s did result in a significant proportion of dwellings being built by industrialized methods, although this would be halved if *in situ* and rationalized traditional construction were excluded from consideration.[66] Detailed research revealed, however, that the way in which low-income accommodation was provided in the UK meant that conditions were not ideal for a self-sustaining industrialized production process: in short, too many authorities were responsible for housing, too many companies responsible for production and too many systems were available, in relation to the necessary concentration of the housing market and the building industry. Between 1965 and 1977 the average size of industrialized building projects was 120 dwellings. In addition, the time-scale of active government support was too short.

In relation to the UK it is worth mentioning that so much attention had been paid during the early 1960s to the future potential of heavy panel systems, especially in the context of high-rise flats, that the decline in the use of heavy panel became synonymous with the decline of industrialized building as a whole.[67] Similarly, the decline in high-rise flats was seen as the decline of industrialized building. Heavy panel and high-rise flats were indeed closely related. However, on the basis of the Department of the Environment's broad definition, industrialized methods continued to be used; *in situ* continued to pay an important role, while reliance upon timber systems increased dramatically. Both of these were at the less sophisticated end of the spectrum of industrialized building and were used for low-rise flats and houses, which have been traditionally favoured by the British. In addition both *in situ* and timber systems could be given a traditional appearance. From their longevity it must be assumed that they also did not need long production runs and were more appropriate to the average size of housing scheme in the UK.

In many ways the UK example of a welfare state economy, while occasionally resulting in high-rise flats, steered a course between very little publicly provided low-income housing on the one hand and an over-abundance of high-rise industrialized accommodation on the other. In the long run, the fragmented nature of the housing system in the UK and consumer preference (more the local authority housing committee than the tenant) influenced the type and form of industrialized building that ultimately prevailed in the UK.

Rhetoric and reality
So far in this section we have mainly considered the extent of actual construction and how it has been affected by the social system in general and the dwelling production system (or lack of one) in particular. We now turn to consideration of the extent to which the implementation fulfilled the expectations aroused by the theory and its rhetoric.

Across the wide range of political and economic systems studied, we found that factory-based industrialized methods of construction, as developed during the 1950s and 1960s, failed to reveal the substantial benefits implied by the rhetoric. The analogy to the assembly line production of motor cars is invalid because the nature of the product, and the whole process of production, are inimical to the creation of a system of machinery for housebuilding. In sum, the product is bulky and heavy and thus unsuitable for transportation; it is constructed of materials ill-suited to achieving the tolerances necessary for technical assembly; it varies both in requirements of size and fittings, which further complicate questions of jointing and standardization; it is very expensive in relation to average earnings, and it has a long life, which make it a peculiar commodity; and there are individual requirements for each structure dictated not only by its prospective inhabitants but by the variable nature of foundation and site conditions, the accompanying services and local building regulations.

When compared not to rhetoric but to traditional building the evidence of cost competitiveness is mixed. In general this may be seen in various UN reports. In the UK, capital-intensive factory-based methods were only economically efficient in the case of high-rise flats. This was by a margin of 3 to 13% (9% average) on tender prices. Tender prices are inadequate as a basis for comparison. But construction of high-rise flats was 35% and 80% more expensive than low-rise dwellings and houses respectively, so highly industrialized methods were only economically competitive for an expensive form of construction.[68] Furthermore, high-rise accommodation built within tight cost restraints is unsuitable as accommodation for low-income families, particularly those with young children.[69] Once large-panel construction is removed then *in situ*, rat-trad (rationalized-traditional) and timber-based systems are also competitive with traditional construction, but these are unsophisticated types of industrialized building and are certainly at odds with the theory.

At the levels of construction achieved in the USSR, overall cost savings amounted to 10% and labour savings from 20 to 25% by comparison with normal brick construction.[70] In view of the severity of the Russian winters, the latter is not to be ignored but neither the total cost savings nor the labour savings are as high as would be expected from the rhetoric. Furthermore, it has been seriously questioned whether these calculations of cost would have fully taken into account the massive investment of capital-intensive methods. In addition the financial commitment to research and development was far greater and probably not considered in the cost comparisons. During the early period of implementation mainly four-to five-storey buildings were built, although in the very large cities seven to eight storeys predominated; later the trend was to higher buildings.[71] It is of interest that

Premier Khrushchev has been specifically identified as being responsible for the original decision to build only four- to five-storey buildings.[72] Despite doubts as to the accuracy of cost data, the scale of production in the USSR is so great that Soviet experience would have to be considered seriously under similar conditions of climatic severity and shortage of skilled labour.

In the USA, Operation Breakthrough failed to make any impact on the market, which suggests that industrialized building was unable to compete with either traditional methods or the mobile home industry.

In the case of Iran the costs of prefabricated construction were prohibitive. Even the estimates made by the UN advisers of the cheapest form of concrete industrialized building, were 6% more expensive than the most widely used method of high-rise construction, namely metal frame and brick infill. In turn metal frame was 57% more expensive than fired brick and metal beam. Actual examples of prefabricated low-rise construction were between 200 and 400% more expensive than the most widely used method of low-rise construction permitted in urban areas, i.e. fired brick and metal beam.[73] The latter was itself 300% more expensive than the current form of really low-income housing prevalent in rural and peri-urban areas, which employed sun-dried mud-brick and mud.

In relation to developing countries in general, authorities such as Terner, Turner and Strassman, and Yeh and Laquian have been even more negative,[74] summarized by Burns and Grebler: 'the experience in LDC's has been almost wholly negative'.[75]

The quality of construction achieved using advanced methods of industrialized building has left much to be desired. In the USSR the low quality of industrialized building has been publicly acknowledged at a senior level within the government itself.[76] In the UK the technical failure at Ronan Point of one of the most widely used systems (Larsen Nielsen) has been compounded by the high maintenance costs of many other advanced systems.[77] As important in relation to quality is the fact that in both the USSR and the UK, advanced methods of industrialized building have only been economically competitive for high-rise construction. But to reiterate, high-rise dwellings built within tight cost restraints are unsuitable for low-income families, particularly those with young children.

PRESENT STATUS OF THE THEORY

The success, failure or mediocrity of industrialized building is of direct relevance to those interested in housing policy. Above we have also illustrated aspects which should interest those concerned with the relationship between technology and society.

In the introduction and the section on initiation of policy, we mentioned that advocacy of industrialized building still took place from time to time. Not only could this result in expenditure of time, effort and money, but it also restricted consideration of possible alternatives. An international lobby in favour of industrialized building could still be identified. The existence of an international network of academics and advisers is one of the means whereby the theory has been kept alive. However, if one compares the original predictions of a twenty- to thirty-fold cost advantage, or even the later 200 to 300% (2–3 times) advantage, with the arguable achievement of only a 10 to 20% improvement in costs for an inadequate form of accommodation, the question arises, how have the experts maintained their faith?

In the face of its repeated failure to fulfil the expectations aroused by the rhetoric, the continued advocacy of industrialized building should prove of particular interest to those interested in the relationship between technology and society.

Modification of theory, maintenance of rhetoric
The way in which commitment to the theory has been maintained is almost Kuhnian. First, attention is always focused upon restatement of the theory. Time and again the basic premises have been restated. It is probably useful to illustrate the phraseology used:

> The very few and scattered data that have been made available concerning the effect of repetition on building costs do not permit a proper analysis and evaluation of the different factors affecting cost savings in repetitive work. However, all information submitted on the subject, as well as hypotheses based on the effect of repetition and operational time, point clearly to the fact that substantial savings in building costs can be gained directly and indirectly from the repetition of identical building operations in long work sequences.[78]

Second, increased stress was laid upon the necessary preconditions to be observed in order for the theory to reveal its benefits: large-scale developments; long production runs; good organization.[79]

Third, the thing to which industrialized or prefabricated building should be compared, changed. Initially, the comparator was the productivity increase achieved in the manufacturing industry. Whilst this *remains* the rhetorical reference, it quite quickly disappeared from any hard and fast discussions. The comparator became the traditional industry.[80] This was probably the most important development, and the question remains, how has the rhetorical power been retained in the face of abject failure by comparison with the original point of reference, despite the lowering of short-term expectation? We will return to this below.

Fourth, stress has been laid upon the numerically successful cases, particularly in Eastern Europe and the Soviet Union, and as we have seen the numerical record in these countries is impressive.

Fifth, instances to the contrary have been explained away: external imperfections, such as restrictive building practices, or the faulty implementation of the theory, have been seen as the major reasons why the benefits have failed to reveal themselves. For example, in the UK the government did not maintain either its high-rise flat subsidy or the industrialized building drive for long enough; the industrial production runs were insufficient for economies of scale to be revealed; too many systems were on the market, many of which were not properly tested. In the USA, by contrast, the major complaints were that the major manufacturing companies had not accepted the challenge and did not enter the housing market, and that the means whereby dwellings were provided were too fragmented.

A variation here would be to state a premise, for example 'the process of improving labour productivity through repetition of a series of identical building operations does not differ from that observed in the manufacturing industries',[81] yet in the next breath to caution 'the main difference as compared with manufacturing lies in the number of adverse factors influencing building work'.[82] In other words, to allow an escape route.

The last example leads into the way in which qualifying clauses were introduced into the discussion. These included transport costs and organization problems. In turn these found their way back into the list of preconditions for success, which enabled the following advice: 'large experience gathered by countries which started earlier with building with large panels should serve the third world countries to plan properly [sic] their progress'.[83]

Sixth, we find an increased insistence upon the importance of other economies. For example, in a 1965 report it was concluded that 'while substantial savings' could be gained directly and indirectly from the repetition of identical building operations, the 'indirect savings gained from a reduction of construction time predominate and are growing in importance'.[84] Elsewhere it was pointed out that the elements of perhaps 25% capital and 50% labour in the production costs of concrete structures had been completely reversed and it stressed that the objective of government measures in this field was '*to promote the production of more dwellings with the same amount of labour and without increase in building cost*'.[85] Similar remarks were made about the speed of production.[86]

Seventh, shortcomings or limitations could be admitted but in such a way as to discount their likely effect. On the one hand it is particularly interesting that both Bowley and White, writing in the early 1960s, tended to highlight those historical developments which were expected to play the greatest part in industrialized housing during the 1960s: multi-storey pre-cast concrete panel in large-scale production.[87] On the other hand, while these studies contained warnings regarding maintenance,[88] cost,[89] methodology of government intervention,[90] the limitations of concentrating mainly upon the outer shell,[91] and the risks involved,[92] they were written in such a way as to play down these factors. They emphasized that the change in circumstances, which had led to the development of large-scale projects of multi-storey construction, had created conditions in which the potential of prefabricated pre-cast concrete could be realized.

Eighth, while the major change has been to alter the point of reference, from the dramatic increases in productivity achieved in the manufacturing industry to a straightforward comparison with the traditional building industry, the ultimate objective remains unchanged, and proponents now speak in terms of 'a gradual process leading to the full production of building'.[93]

> We should not identify industrialization solely with the latest technological achievements displayed by completely prefabricated buildings.

> This is the target, undoubtedly, but in reaching that stage of development a number of gradual technological changes are indispensable, starting with rationalization, followed by partial prefabrication, and then by complete prefabrication.[94]

Such step-by-step development has been frequently reiterated. For example, in the report on the 1975 UN Interregional Seminar on Design and Technology for Low-cost Housing, we read that 'the introduction of industrial methods in construction in developing countries is also inevitable but must be carried out gradually and carefully'.[95] The various stages of the process of industrialization of building were considered to be: (i) conventional construction methods; (ii) rationalized methods; (iii) partial prefabrication; and (iv) complete prefabrication.[96] In addition, it was pointed out that the most advanced methods should be reserved for use in large settlements.

It may be seen that while the ultimate objective remains essentially the same as it did for the original advocates, the path towards it has become somewhat more tortuous: it must be approached carefully and in stages. This is a further development of the most important internal theoretical modification: while introductory discussions have always focused upon the radical increases in labour productivity, the actual yardstick in practice has been whether it could be competitive with traditional methods. However, the retention of the ultimate objective—complete prefabrication—permits retention of the rhetoric.

So far we have limited ourselves to discussing aspects of the theory itself. But throughout the discussions about industrialized building there was an additional factor: industrialized building represented 'progress'. As mentioned above, in 1963 at a conference on industrialized building, one leading contractor was reported as saying that 'industrialized building meant building in the manner of other things in the 20th Century'.[97] Industrialized building was 'inevitable',[98] an 'unmistakeable and irreversible trend led to the transfer of site work to factories'.[99] An often repeated phrase was the 'impossibility of putting the clock back'.

The corollary was the general view that the ordinary methods of construction were backward. A UN report was adamant that

> the special problems of applying industrial principles of production in building should, however, not be exaggerated and should not be allowed to excuse the use of obsolete techniques and out-of-date methods of planning and organization.[100]

Traditional building consisted of 'out of date work under primitive conditions'. In 1966 George Lothian, the General Secretary of the Amalgamated Union of Building Trade Workers (UK), stated 'the building industry is not making as efficient a contribution to the nation's welfare as it should' and 'too many firms are still in the "wheelbarrow and spade" age'.[101]

How can this rhetoric have remained unchanged in the face of its inability to fulfil the claims originally made on its behalf? Indeed, certain proponents of industrialized building have long been aware of the difficulties at both the theoretical and practical levels. We have seen the rationalizations and modifications that have been made to the theory and the way in which the ideal has been retained, albeit at a distance. Yet the phraseology in the literature suggests that one of the major factors which enabled the proponents to maintain their conviction in the face of contrary evidence was the concept of a technology which was consonant with other aspects of a modern world, and was not backward or primitive like traditional building.

Taken together these factors mean that the theory still has a powerful impact, especially upon those coming into contact with it for the first time. In the past, the application of the machine to many types of work has greatly increased productivity, and significant economies of scale have been revealed through mass production. As Bemis put it, 'why not housing?'. The promise of increased production of a higher-quality modern product at a lower cost, combined with ostensibly pleasanter working conditions for the workers, is almost irresistible. The theory has been modified so that its anticipated benefits are far lower than originally mooted, yet the ultimate objective has been maintained, allowing the rhetoric to retain its persuasive force. This rhetoric resonates with the ethos of progress and an image of a modern society, and has additional strength by contrasting so strongly with aspects of traditional building that seem to be primitive and backward.

Continued propagation of the theory

There are professional, institutional, entrepreneurial and psychological factors which have supported the continued propagation of the theory. These may be summarized as follows:

First, it is a compelling theory which has a powerful effect upon those who come into contact with it for the first time. In addition the theory is usually discussed in the following context: on the demand side the scale of shortage is always described as extremely high (at least an order of magnitude higher than is possible to counter using existing methods) and the solution is required immediately in the face of an absence of skilled workers: a miracle is required! The rhetoric surrounding industrialized methods is particularly vibrant under these circumstances: 'revolutionary' means are called for. Further, from the point of view of the decision maker, there is the apparent ease with which the problem can be solved given the existence of solutions in the industrialized world. In this respect the sheer numbers of dwellings built by industrialized methods in social welfare and, particularly, in socialist countries are impressive. The temptation for the capital-rich developing countries to use these methods to augment their housing stock is real.

Second, senior members of the design professions who advise governments and international agencies on housing policy remain committed to the theory and there are social and institutional networks through which they can operate.

Third, there are companies that developed systems during the 1950s and 1960s which continue to sell their wares. Nearly every company that had a system visited Iran during the course of its Crash Housing Programme (1975-8). Later reports indicated growth areas in China, the Middle East, Malaysia and Indonesia. This activity is not the dominant one during the initial stages. The decision to follow such a course of action has first to be taken by the government, which is generally acting on the advice of its technical officers from building research centres or academics in the universities. These people are linked into the international professional networks mentioned above. For example, the Director of the Building and Housing Research Centre in Tehran, Dr Neghabhat, was Iran's representative on the CIB, and he was a board member of the International Association for Housing Science.

Fourth, not only companies but also some governments are eager to provide hardware; unlike companies, governments are also interested in providing software in the form of expert advice. Both socialist and welfare states have played an active part in the development of industrialized methods. In Denmark the government played an important role in the development of the Larsen Nielsen system and actively continued to promote its use.

Finally, a major contention of this paper is that these factors were bound together by a common desire to be progressive and modern, and by an antipathy towards what was perceived as the inadequacy and backwardness of traditional methods.

It is at this point that I consider the effect of the theory to have been most damaging, because in both industrialized and developing countries it has obscured serious consideration of possible alternatives, in particular the need to create building skills.

CONCLUSIONS

Over the past forty years there have been many calls upon modern technology in general to play a greater part in relieving the housing shortage. One particular form, industrialized building, was at one stage regarded as the panacea which would drive down costs. Whilst this has been seriously challenged by alternative views, it continues to find resonance amongst housing authorities in various parts of the world. It should be useful, therefore, to examine the subject from the point of view of housing policy as even initiation of policy has serious short-term costs in time, effort and money and obscures consideration of possible alternatives. Such an examination also reveals manifold interactions between technology and society.

Theory

The theory was quite simply based upon an analogy to the mass production of motor cars, incorporating science and technology at every stage. If the process of producing dwellings were modelled upon that of motor cars it would result in greater volumes of a better-quality modern product at a cheaper cost and without the need to increase scarce skilled building resources. The language used in formulation and discussion of the theory was redolent with references to the inevitable benefits which would result from the application of science and technology to the dwelling production process. Probably the most important aspect was the rhetoric that accompanied the theory. It spoke of phenomenal increases in output, speed of production and quality for a much lower outlay of resources. The theory was first formulated during the 1920s. It has had several periods of dominance (and although not as prevalent today as in the 1960s and 1970s, it still persists). On one level its longevity substantiates the contention that the theory is a reflection of an ethos which is progressive and imbued with positive expectations about science and modern technology. In 1965 White remarked that:

> Prefabrication seems to have exerted a perennial fascination on industrially minded architects and others searching for the magic key that would unlock the factory that was going to produce the 'house of the century'. Houses that could be produced and marketed like consumer goods and whose price would reflect the economies attributed to mass production.[102]

An important corollary of the consonance between the theory and a prevailing ethos was an aversion to anything backward and primitive, both being attributes of traditional housing and its process of production.

Propagation
Propagation of the theory was mainly carried out by members of the design professions, who were able to maintain close contact with one another through various professional networks. These networks included amongst their members senior academics, directors of national institutes of building and architecture, senior government officials responsible for making housing policy, senior officials who carried the housing brief for international agencies and representatives from the construction industry. In some cases the roles became interchangeable.

Initiation of policy was facilitated by these forums but in four detailed case studies it was clear that an entrepreneurial-type activity was required on the part of the technical advisers combined with access to a position of power *vis-à-vis* policy making.

Implementation
The building of low-income accommodation has only been extensive in industrialized countries with significant public housing programmes. The larger the public housing programme the greater the use of industrialized building and the greater the proportion of sophisticated methods. In capitalist market economies and in developing countries very little use has been made of industrialized methods. In no case were factory-based methods first used to provide accommodation in the private sector.

In relation to socialist countries the implementation of industrialized methods has often been assumed to be unproblematic. Such a view is simplistic. In the USSR not only is the housing industry more concentrated and more amenable to control than in welfare economies, but also capital investment in factories and production lines has been far greater. Even so, the following indicates that industrialized methods do not simply flow from a particular social system. Firstly, concentration of the industry resulted from deliberate action over a long period of time; it did not just occur spontaneously. Secondly, it has taken much longer than expected to institute industrialized methods. In 1959 it was predicted that by 1964 large panel methods would account for the majority of new dwellings, yet these methods only accounted for 50% of new urban housing by the mid-1970s. Further, the type of prefabrication which in 1959 was thought to hold the most promise—multi-storey 'box' construction—still remains on an experimental basis. Other problematic matters (cost, quality, type of building) will be mentioned below.

Even in industrialized countries with significant housing programmes an internal commitment within government is necessary for implementation over a long period of time. Thus, in both the USSR and the UK the initiation of policies stemmed from activity by technical advisers, but the widespread adoption required committed activity on the part of influential individuals and groups within government. In both the USA and Iran the initiation of policy stemmed from technical advisers, but the lack of an institutional form of government commitment as embodied in a public housing sector meant industrialized methods did not become widespread.

Long-term implementation is further affected by a combination of political, technical and economic factors. In capitalist economies the nature of the housing process is too diverse, the housing-building industry is insufficiently capital-intensive, and the level of government control is too weak for the methods to take root. Further, adverse consumer reaction to costs and type of dwelling is rapid. In welfare economies that portion of the building industry dealing with public housing is relatively concentrated and the housing process is subject to a measure of control. Thus, with specific government commitment the policies may be implemented over a longer period of time. In welfare states it appears that the combined effect of cost, technical adequacy and consumer reaction take about a decade to be appreciated. In socialist economies the industry is concentrated and the means of control are tighter. In addition consumer reaction takes still longer to surface.

It is quite clear, therefore, that widespread implementation, over a long period of time, of policies on industrialized building, is heavily influenced by the social framework within which the policy has to operate.

Comparison of results with theory
From the case studies it was found that factory-based industrialized methods of construction, as developed during the 1950s and 1960s, definitely failed to reveal the considerable benefits implied by the analogy to the assembly-line production of motor cars.

When compared, not to the rhetoric but to traditional building, the evidence of cost-competitiveness is mixed. In the UK advanced capital-intensive methods were only economically efficient in the case of high-rise flats. They showed an average advantage of 9%, based on tender prices, but in the context of a type of accommodation which is 35% and 80% more expensive than low-rise flats and houses respectively. Simpler systems were able to compete for low-rise flats and houses. In the USA, Operation Breakthrough failed to make any impact on the market, which suggests that the methods were unable to compete. In Iran the costs of prefabricated construction were prohibitive.

In the USSR industrialized methods have only been used for buildings over four storeys high. While overall cost savings amount to 10% by comparison with normal brick building, it may be seriously questioned whether these calculations of cost have taken into account the massive investment in capital-intensive methods. However, given the length and severity of Russian winters, factory production is very important because it extends the building season considerably.

The quality of construction achieved using advanced methods has left much to be desired. This is certainly at odds with the theory. In the USSR the low quality has been publicly acknowledged at senior government levels. In the UK the technical failure of the Larsen Nielsen system (at Ronan Point) has been compounded by the high maintenance costs of many other advanced systems. As important, in relation to quality, the type of dwelling which resulted from the use of advanced industrialized systems was generally the high-rise flat, which is another example of the severe limitation of theory. Under conditions of cost restraint high-rise flats are seldom appropriate for the accommodation of low-income families, particularly those with young children.

Professor Marion Bowley referred to local authority housing between 1919 and 1939 as 'perhaps the outstanding peace time experiment in state intervention'.[103] The literature on the post-war non-traditional housing revealed that it too was an experiment.[104] The use of multi-storey industrialized systems during the 1960s and 1970s may also be seen as an extended experiment which was carried out on local authority tenants.[105] In the USSR this whole approach has been viewed somewhat naively in glowing terms by an official publication during the 1960s:

> Moscow, Leningrad, Kiev and other Soviet cities are not the only 'experimental laboratories' of architects and builders; housing research has swept the entire country, has come to every town and village.[106]

In the UK the types of industrialized building which ultimately prevailed—in terms of technical soundness, cost and consumer (i.e. housing committee) preference—were extremely unsophisticated, in particular *in-situ* concrete and timber systems. The predominant system 'No-fines' *in situ*, was developed after the First World War, while the timber systems of the 1970s were considered by the developer of the most widely used timber system to be retrogressive by comparison with those of the 1960s. Both *in-situ* and timber-based systems were particularly suitable to low-rise flats and houses, which are the types of accommodation preferred by the British, and these industrialized systems can be given an outward appearance that is indistinguishable from traditional brick dwellings.[107]

While theory, even fashion, may be an important factor in the initiation of public sector programmes—during which time a great deal of time, money and effort may be expended—in the long term structural features of the society determine the extent to which the theory will be implemented, and social factors will influence its manifestation. This study strongly supports contentions made by Pavitt and Walker in their survey of government policies towards innovation. In relation to public services they considered that assessment was far from perfect and those making decisions are often 'not those who eventually have to issue the equipment, and to answer for its cost effectiveness'.[108]

In the context of public sector policy in general, it is important that the research was able to identify the existence of an international body of influential technical advisers who were committed to the propagation and implementation of the theory. Initiation of policy in the public sector can often involve great expenditures of time, money and effort. Public justification for the theory is often gained through the rhetoric. In this case the results certainly did not live up to the claims implied by the rhetoric. Even by comparison with traditional building, several important social preconditions must exist before implementation of the theory can be said to compete marginally in terms of cost. More importantly, in socialist and welfare economies the results of these extended experiments may take over a generation to work themselves through the various stages of: (a) implementation; (b) emergence of shortcomings; (c) acknowledgement of shortcomings; and (d) corrective action. The latter applies both to the unsoundness of much of the technology,[109] and to the inappropriateness of the type of accommodation provided (under conditions of cost restraint).

Perhaps truly modern inventions such as superconductors, optical fibres and lasers may be able to act for a time independently of social constraints. However, in such a basic and technologically humble field as housing, developments are far more subject to social factors.

Present status of the theory
I have demonstrated the means whereby the faith has been maintained in the face of evidence to the contrary, the most important point being the change in the point of reference. Traditional housing—instead of the results of mass production in other industries—became the datum. In addition the ultimate objective—complete prefabrication—was retained (and thus the right to slip into the rhetoric), but several stages had to be completed in the journey to the eventual goal. (When used in the context of labour-abundant and capital-scarce economies, the retention of complete prefabrication as the ultimate form of dwelling is probably the most important aspect of recent developments; we will return to this below.) Various international forums existed which enabled contact between directors of national building institutes, senior academics, representatives from the construction industry and housing policy makers from governments and international agencies. Some governments and contractors still worked to market their dormant capacities.

Four elements of the assessment as to the need for industrialized methods were common to all four case studies. They were also common to the deliberations on this topic by the CIB and the United Nations. The first was the magnitude of the housing shortage; the second was the shortage of existing building skills; and the third was the need for a swift solution to the housing problem. The fourth was the consonance between the theory of industrialized building and the policy makers' idea of what constituted a progressive, modern and science-based technology in keeping with the way other things were done in the twentieth century. A corollary of the latter was the rejection of traditional methods and materials of construction as inadequate and backward and not in keeping with the modern world. The impact upon first-time hearers remains powerful.

It is clear from the Iranian case that, in the mid-1970s, 'industrialized building' could still evoke policies aimed at a radical improvement in housing conditions, in countries which are neither industrialized nor have large housing programmes. The combination of factors summarized above could be brought to bear under inappropriate economic, social and technical conditions. This had the immediate result of a vain expenditure of a great deal of time, effort and money: yet another failed experiment in public housing policy.

Concentration upon industrialized building obscured serious consideration of possible alternatives, such as 'self-help'[110] or increasing the supply and quality of craft skills. In connection with the latter, the literature provides a prime example both of the ethos of the time and the way in which possible alternatives were marginalized. In the mid-1960s a study of operative skills was carried out in the UK at the Building Research Station, as a response to the National Joint Council for the Building Industry (which represented both employers and unions). The summary of the proposed research began with the sentence that

> it was no longer appropriate to consider the industry to be made up of craftsmen on the one hand and unskilled labourers on the other

and it concluded with the need for

> an early contribution . . . to the transition of the building industry from a craft to an industrialized basis.[111]

The option of increasing the supply and quality of craft skills was not seriously considered, even though this option had been taken during the 1920s in the UK. In 1945 Bowley argued that the rapid increase in the number of houses built between 1924 and 1934 had been 'made possible by the increase in the number of skilled men'.[112] The increase had been initiated in 1924 when the building unions 'relaxed their rules so that the number of skilled men could be increased'. This was the 'most essential step in any attack on the war-created shortage of housing'.[113]

This is not to say that a housing shortage could be solved overnight by an increase in supply and quality of craft skills—a solution to the housing shortage requires consistent political commitment and the investment of considerable physical resources. Further, the means of increasing the supply and quality of craft skills would require careful consideration. But progress along these lines would mean an increase in the resource base necessary to provide the type of housing preferred by the majority of people:

technologically sound, weather-resistant, low-rise accommodation. The current climate in the industrialized world might make policy makers more receptive to these ideas. However, the study has shown that for action to be taken not only must an influential body of people begin to promote the idea, but also strategies of implementation should be sensitive to the full complexity of local conditions.

We have seen that concentration upon industrialized building made people look for solutions in a particular direction which then meant that other options were not even taken seriously. In this sense, for labour-abundant and capital-scarce economies, retention of the ultimate objective of complete prefabrication and the stages whereby progress towards that goal may be achieved has been a serious obstacle to consideration of government-sponsored alternatives more in keeping with socio-economic conditions. And the institutional networks still exist through which the modified theory may be fostered.

SUMMARY

Since the Second World War there has been an acknowledged shortage of adequate housing for low-income groups in both industrialized and developing countries. While various solutions have been proposed, may authorities have argued that mass production of housing using prefabricated or industrialized methods of building would solve the housing crisis. It was explicitly stated that the industrialization of building would obtain results which would be orders of magnitude better than those obtained using traditional methods. Policies based on these principles have been adopted by countries which occupy very different positions in the socio-political spectrum. In many countries a great deal of housing has been built using industrialized methods.

Besides the direct implications for housing policy, there are several aspects of this phenomenon which are pertinent to the relationship between technology and society. Firstly, the way in which the theoretical concepts were influenced, generally by an ethos of the time which was pro-science and modern technology, and, specifically, by the concept of mass production as applied to motor cars. A corollary was the rejection of traditional methods of building as inadequate and backward.

Secondly, the design professionals played a crucial role in the formulation of policy and the initiation of programmes. Here it is particularly interesting that the theory transcended political and economic differences; in part this was due to the existence of international forums for formal and informal communication. The individual initiatives that had to be taken in order to have the theory taken seriously at senior policy-making levels is of particular interest to the internalist/externalist debate. The influence of technical advisers is also important when one considers that a great deal of time, effort and money may be involved in the initiation of public policy, irrespective of the outcome. An unfortunate side-effect of such concentration is that it obscured serious consideration of possible alternatives.

Thirdly, a striking example of the effect of the nature of society upon the development of a technology is the extent to which the use of industrialized building has depended upon the existence of public sector housing. The greater the extent of public sector housing the greater the reliance upon more sophisticated methods.

Fourthly, implementation of policies based on the theory of industrialized building failed to live up to the order of magnitude of the benefits anticipated by the theory (proclaimed by the accompanying rhetoric) with respect to quality, cost, time and labour saving. Furthermore, the type of structure most suited to the advanced systems—high-rise flats (which were much vaunted during the early 1950s and 1960s)— were unsuitable as accommodation for low-income families, particularly those with young children, especially if the dwellings were built within tight cost constraints.

Even by comparison with traditional methods of building, evidence of cost saving is mixed. Despite the dissonance between theory and practice—the manifest failure when compared to the rhetoric, only mixed evidence when compared to traditional building—industrialized building continues to be advocated by its protagonists. Thus, a fifth point of interest to those concerned with the links between technology and society would be the way in which the theory and the accompanying discussion have altered over time in the almost Kuhnian manner.

Finally, despite the limitations of the theory, the rhetoric remained almost entirely unscathed. It is particularly evocative when used on a first-time audience—especially one made up of inexperienced housing authorities confronted by an urgent large-scale housing problem which apparently cannot be solved quickly using existing resources. When such a situation arises there is still an international network prepared to advise upon implementation. Given the numbers of dwellings that have been built in some countries by industrialized methods this should not be surprising, particularly in view of the way in which the theory has been modified over time: while the ultimate objective remains the same, the definition has been expanded to include less sophisticated methods of building. As indicated above, concentration upon policies of this nature not only absorbs resources but tends to obscure serious consideration of possible alternatives; this is particularly disturbing in the context of labour-abundant capital-scarce economies.

Acknowledgement

I would like to thank Professor Christopher Freeman, Science Policy Research Unit, University of Sussex, for his advice and encouragement.

Footnotes and References

1. Mario Piche, 'International year of shelter', *International Journal for Housing Sciene and Its Application*, Vol. 11 (i), 1987, p. 1. See also S. Chameki, 'The world housing problem', in D. Sfintesco (ed.), *Conference 2001: Urban Space for Life and Work*, Paris, UNESCO/Council of Tall Buildings and Urban Habitat, 1977, Vol. 2, p. 343; and S. Chameki, 'The world housing problem in figures', *International Journal for Housing Science and Its Application*, Vol. 1 (1), 1977, p. 3.

2. Charles Abrams, *Man's Struggle in an Urbanizing World*, Cambridge, Mass., MIT Press, 1964, p. vi. D.V. Donnison, *The Government of Housing*,

Harmondsworth, Penguin, 1967, p. 9. L.S. Burns and L. Grebler, *The Housing of Nations, Analysis and Policy in a Comparative Framework*, London: Macmillan, 1977, p. 230. O.F. Grimes, *Housing for Low-Income Urban Families: Economics and Policy in the Developing World*, Baltimore and London: Johns Hopkins University Press for the World Bank, 1976, p. 3: 'housing was looked upon primarily as a physical phenomenon'.

3. Grimes, op. cit. (see note 2).

4. UNECE, *The Major Long Term Problems of Government Housing and Related Policies* (ST/ECE/HOU/20), New York: UN, 1966, Vol. I, p. 162; see also CIB (ed.), *Towards Industrialized Building: Proceedings of the Third CIB Congress 1965*, Amsterdam, New York: Elsevier, 1966, p. v; F. Lea, Director of the Building Research Station (BRS) and President of the CIB, 'Closing Address' in CIB (ed.), *Innovation in Building—Proceedings of the Second CIB Congress, Cambridge, 1962*, London: Elsevier, 1963, Supplement, p. 63.

5. M. Bowley, *The British Building Industry: Four Studies in Response and Resistance to Change*, Cambridge: CUP, 1966, p. 291.

6. Burns and Grebler, op. cit., p. 230 (see note 2).

7. R.T. McCutcheon, 'Technical change and social need: the case of high flats', *Research Policy*, Vol. 4, 1975, 262–289.

8. Idem, 'Highrise flats in the UK and Iran: a comparative review', paper presented at *Conference 2001: Urban Space for Life and Work, UNESCO, Paris, 21–25 November 1977*, Hamadan: Bin Ali Sina University, Mimeo, October 1977, 56 pp. Also summarized, 7 pp.

9. The definition of prefabrication and industrialized building has been discussed at length in R. McCutcheon, 'Modern construction technology in low-income housing policy: the case of industrialised building', University of Sussex: D.Phil. Thesis, 1979, Chapter 2: pp. 40–64.

10. Le Corbusier, *Towards a New Architecture*, London: Architectural Press, 1927; see also Le Corbusier, *The Radiant City*, (1933), New York: The Orion Press, 1964; and Le Corbusier, *The Marseilles Block*, London: The Harvill Press, 1953.

11. Walter Gropius, 'Industrialised building', *Constructor*, August 1969, Vol. 17; originally published in 1910. See also W. Gropius, *The New Architecture and the Bauhaus*, London: Faber & Faber, 1936, pp. 25–6.

12. Guy Oddie, 'The New English Humanism: Prefabrication in its Social Context', *Architectural Review*, September 1963, Vol. 43(9), pp. 180–182.

13. Albert Farwell Bemis, *The Evolving House*, Vols 1 to 3, Cambridge, Mass: The Technology Press, MIT, 1933 to 1936. Foreword to all volumes, pp. vii and viii.

14. Ibid. The Tudor Walters Committee, which was formed in 1917 'to report on methods of securing economy and despatch in the provision of [working class] dwellings', reached certain conclusions pertinent to our study which have been reiterated ever since. On the subject of standardization, the committee found that 'to a great many witnesses', this meant the manufacture of complete houses of two or three standard patterns, on the lines adopted for the manufacture of the motor car. Other witnesses considered that standardization was limited to components such as doors, windows, fireplaces and the like. Such standardization would enable large-scale production, which would lead to cost economies. The committee reported not only that large firms which had been carrying out wartime contracts for the government had depleted the staffs of the small house-builders scattered up and down the country, but that only these large firms of civil engineering contractors would be immediately capable of handling large housing schemes. Local Government Board, *Report of the Tudor Walters Committee*, London: HMSO, October 1918; quoted in R.B. White, *Prefabrication: A History of Its Development in Great Britain*; National Building Studies Special Report 36, London: HMSO, 1965, pp. 42–43.

15. The Committee for the Industrial and Scientific Provision of Housing, *Housing Production or the Application of Quantity Production Technique to Building, Its Social, Commercial and Technical Possibilities and Requirements; First Report*, London: 3 Albemarle St., Jan. 1943, p. 1.

16. Ibid. p. 6; see also pp. 12 and 15.

17. Le Corbusier, 1927, op. cit. and Gropius, 1936, op. cit., pp. 25-26.

18. Henry Ford, *Encyclopedia Britannica*, 1953, Vol. 15, pp. 38, 39.

19. Bemis, op. cit. Vol. 2, p. 227.

20. The Committee for the Industrial and Scientific Provision of Housing, op. cit., p. 6; see also pp. 12 and 15.

21. R.B. White, op. cit., p. 3.

22. UNECE, *Proceedings of the Seminar on Changes in the Structure of the Building Industry Necessary to Improve Its Efficiency and to Increase Its Output. Prague, Czechoslovakia, 19-30 April 1964* (ST/ECE/HOU/13), New York: UN, 1965.

23. UNECE, *Effect of Repetition on Building Operations*, (ST/ECE/HOU/14) New York: UN, 1965, p. 1. See also UNIDO, *Prefabrication in Africa and the Middle East* (UNIDO/ID/100), Vienna: UNIDO, February 1973, p. 10.

24. UNECE, *The Future Design, Production and Use of Industrially Made Building Components; Report on the Proceedings of the Second ECE Seminar on the Building Industry, Paris, 24-29 April 1967* (ST/ECE/HOU 36), New York: UN, 1969, Vol. 1, p. 69.

25. UNECE, *The Major Long Term Problems of Government Housing and Related Policies* (ST/ECE/HOU/20), New York: UN, 1966, Vol. 1, p. 196.

26. UNIDO, *Industrialization of Developing Countries: Problems and Prospects: Construction Industry* (ID/40/2), New York: UN, 1969.

27. I.D. Terner and J.F.C. Turner, *Industrialized Housing: The Opportunity and the Problem in Developing Areas; Ideas and Methods Exchange No 66*, Washington: Department of Housing and Urban Development, January 1972, I-1.

28. UNECE, *Economic Survey of Europe in 1976, Part II: The five-year plans for 1978-80 in Eastern Europe and the Soviet Union*, New York: UN, 1977, p. 99.

29. Building Research Centre, Scientific Research Foundation (Iraq), *Proceedings of the Symposium on System Building and Prefabrication, 1-4 March 1976, Baghdad*, Baghdad: Building Research Centre, 1976, p. vi. Note also the option foregone.

30. Hans Steenfos, 'The basic conditions for introducing building methods in the Third World countries', *International Journal for Housing Science and Its Application*, Vol. 1 (1), August 1977, p. 41.

31. B. Lewicki, 'The economy of large panel systems', ibid., p. 95.

32. Steenfos, op. cit., pp. 41 and 47. See also K. Tharmaratnam and Saw Choo Ban, 'Industrialized housing in Singapore', *International Journal for Housing Science and Its Application*, Vol. 9 (4), 1985, pp. 291-301. Dr Tharmaratnam is an Associate Professor in the Department of Building Science at the National University of Singapore. Dr Saw Choo Ban is the Chief Structural Engineer, Housing and Development Board, Singapore.

33. UNECE (HOU/14) 1965, op. cit., pp. 13 and 15. G. Schindler, 'Report on possible advantages of prefabrication for standardized construction of dwelling houses', Geneva: UN, Jan. 1950; referred to in UNECE, *European Housing Progress and Policies in 1953* (E/ECE/189), Geneva: UN, 1954. For further comments on this aspect see White, op. cit. and L. Mumford, *The Culture of Cities* (1935), London: Secker and Warburg, 1946, pp. 467-469. Ministry of Works, *New Methods of House Construction; National Building Studies Special Report No. 4* (First Report), London: HMSO, 1948; *Special Report No. 10* (Second Report), London: HMSO, 1948; White, op. cit., pp. 155-156. UNECE (HOU/20), 1966, op. cit., pp. 164, 165 and

170. ILO, *Social Aspects of Prefabrication in the Construction Industry*, Geneva: ILO, 1968, p. 38.

34. UNECE, *Government Policies and the Cost of Building* (E/ECE/364), Geneva: UN, 1959, p. 17.

35. Bowley, op. cit., p. 316.

36. R. McCutcheon, 'Industrialized house building 1956-1976', *Habitat International*, Vol. 12(1), 1988, p. 95.

37. UNECE, *European Housing Progress and Policies in 1953* (E/ECE/189), Geneva: UN, 1954; *Proceedings of the Ad Hoc meeting on Standardization and Modular Coordination in Building* (E/ECE/361-E/ECE/HOU/85), Geneva: UN, 1959; *Housing Costs in European Countries* (ST/ECE/HOU/8), Geneva: UN, 1963; *Cost Repetition Maintenance, Related Aspects of Building Prices* (ST/ECE/HOU/17), Geneva: UN, 1963.

38. UNECE, *Proceedings of the Seminar on Changes in the Structure of the Building Industry Necessary to Improve Its Efficiency and to Increase Its Output, Prague, Czechoslovakia, 19-30 April 1964* (ST/ECE/HOU/13), New York: UN, 1965, Vols I, II and III; *Effect of Repetition on Building Operations and Processes on Site* (ST/ECE/HOU/14), New York: UN, 1965; *The Future Design, Production and Use of Industrially Made Building Components; Report on the Proceedings of the Second ECE Seminar on the Building Industry, Paris, 24-29 April 1967* (ST/ECE/HOU/36), New York: UN, 1969, Vols I and II; *Proceedings of the Third ECE Seminar on the Building Industry, Moscow, USSR, October 1970* (ST/ECE/HOU/42), New York: UN, 1971, Vols I and II; *Fourth Seminar on the Building Industry, London, 8-20 October 1973* (ECE/HBP/SEM3), New York: UN, 1973; *Dimensional Coordination in Building* (ECE/HBP/6), Geneva: UN 1974. See also, United Nations Economic and Social Council (UNESC) Committee on Housing Building and Planning, *World Housing Conditions and Estimated Housing Requirements: Prepared by the Secretariat* (E/C.6/13), New York: UN, 1963; *The Development of the Construction and Building Materials Industries with Special Reference to Industrialization: Note by the Secretary General* (E/C.6/16), New York: UN, 1964; *Industrialization of Building: Report of the Secretary General* (E/C.6/36 and Add. 1-9), New York: UN, 1965; *Industrialization of Building: Note by the Secretary General* (E/C.6/52/Add. 3), New York: UN, 1966; *Industrialization of Building* (E/C.6/70 and Add. 1), New York: UN, 1967; *Trends in the Industrialization of Building* (ST/SOA/102), New York: UN, 1970. UN Centre for Housing Building and Planning and Office of Technical Cooperation, *Report of the Interregional Seminar on Design and Technology for Low-cost Housing, Budapest, Hungary, 9-20 April 1974*, New York: UN, 1975; draft; the Conference papers were numbered ESA/HBP/AC.13/1 et seq.

39. UN Economic Commission for Africa, *Prefabrication in Africa: Status and Problems* (E/CN.14/HOU/101), UN, 1973. UN Economic Commission for Asia and the Far East (UNECAFE), *Industrialization of Housing in Asia and the Far East: Trends and Prospects; Proceedings and Selected Documentation of the Seminar on the Industrialization of housing in Asia and the Far East, Copenhagen, Denmark, 26 August-14 September 1968* (E/CN.11/1054), New York: UN, 1972; *Roving Seminar on Standardization and Modular Coordination in the Building Industry in Asia and the Far East, 27 October-6 December 1969. Report of the Group of Experts, 7 January 1970* (E/CN.11/1 and NR/SSMC/L.I); *Standardization and Modular Coordination in the Building Industry in Asia and the Far East. Second Roving Seminar, organized by the ECAFE in collaboration with the UNOTC and the Governments of Denmark, Indonesia, Philippines and Thailand, 6-31 March 1972, Bandung, Manila and Bangkok* (E/CN.11/1 and H/RSSMC/L.I) UN Economic and Social Council, ECAFE: 28 June 1972. UNECLA, *Latin America Seminar on Prefabrication of Houses Sponsored by the United Nations and the Government of Denmark, 13 August to 1 September 1967* (ST/ECLA/ Conf 27).

40. UNIDO, *Industrialization of Developing Countries: Problems and Prospects. Con-*

struction Industry/UNIDO Monograph on Industrial Development (ID/40/2), New York: UN, 1969; *Prefabrication in Africa and the Middle East: Report of the Expert Group Meeting Held at Budapest, Hungary, and Bucharest, Romania, 17–29 April 1972* (ID/100) Vienna: UNIDO, 21 July 1972; *The Construction Industry in Developing Countries: Report of an Expert Group Meeting, Vienna 29 October–2 November 1973* (ID/111), Vienna: UNIDO, December 1973.

41. CIB (International Council for Building Research Studies and Documentation) (ed.), *Building Research and Documentation: Proceedings of the 1st CIB Congress, Rotterdam, 1959*, Amsterdam: Elsevier, 1961; *Innovation in Building: Proceedings of the Second CIB Congress, Cambridge 1962*, London: Elsevier, 1963, and Supplement; *Towards Industrialised Building: Proceedings of the Third CIB Congress, 1965*, Amsterdam, New York: Elsevier, 1966; *Building Cost and Quality: Proceedings of the Fourth CIB Congress, Ottawa, Canada, and Washington, D.C., 1968*, Rotterdam: CIB, 1969; *Research into Practice: The Challenge of Application. 5th Congress, Paris, Versailles, June 1971*, Paris: CIB, 1971.

42. D. Sfintesco (ed.), *Conference 2001: Urban Space for Life and Work*, Paris: UNESCO/Council on Tall Buildings and Urban Habitat, 1977, Vols I and II. Council on Tall Buildings and Urban Habitat, *Monograph on the Design of Tall Buildings* (5 Vols), New York: American Society of Civil Engineers, 1978.

43. International Association of Housing Science and its Applications (IAHS), Oktay Ural *et al.* (eds), *First International Symposium in Low-Cost Housing Problems Related to Urban Renewal and Development, October 1970*, Rolla, Missouri, University of Missouri at Rolla, 1970; Oktay Ural (ed.), *Second International Symposium on Low-Cost Housing Problems, April 1972*, Springfield: National Technical Information Service, 1972; Oktay Ural (ed.), *Third International Symposium on Low-Cost Housing Problems, May 1974*, Montreal: Concordia University; Oktay Ural (ed.), *New Trends on Lower Cost Housing Production Emphasizing the Problems of Emerging Countries* Miami: IAHS, 1975; F. Rad Parvis *et al.* (eds), *Fourth International Symposium on Housing Problems, May 1976*, Elmsford, NY: Pergamon Press, 1976; Adel Fareed, Mostafa El-Hifnawi and Oktay Ural (eds), *IAHS Cairo Workshop on Evaluation of Industrialized Housing Systems, November 1976*, Cairo: IAHS, 1976; V.S. Parameswaran (ed.), *International Seminar on Low-Cost Housing, January 1977*, Madras: IAHS, 1977; R.P. Pama, S. Angel and J.H. De Goede (eds), *International Conference on Low-Income Housing Technology and Policy, June 1977*, Bangkok: Asian Institute of Technology, 1977; Oktay Ural and Aliye Celik (eds), *International Conference on Disaster Area Housing, September 1977*, Ankara: Turkish Building Research Institute, 1977; Oktay Ural and Bulent Tokman (eds), *RCD Lower Cost Housing Workshop, November 1977*, Ankara: Turkish Building Industry, 1977; F. Dakhil, Oktay Ural and M.F. Tewfik (eds), *Housing Problems in Developing Countries: Proceedings of IAHS International Conference in Saudi Arabia, 1978*, Chichester: John Wiley, 1978; Oktay Ural (ed.), *Housing: Planning, Financing, and Construction. Proceedings of the IAHS International Conference in Miami Beach, Florida, December 1979*, Elmsford, NY: Pergamon Press, 1979; Oktay Ural (ed.), *Energy Resources and Conservation Related to Built Environment. Proceedings of IAHS International Conference in Miami Beach, Florida, December 1980*, Elmsford, NY: Pergamon Press, 1980; Oktay Ural and Robert Krapfenbauer (eds), *Housing: The Impact of Economy and Technology: Proceedings of the IAHS International Conference in Vienna, Austria, November 1981*.

44. McCutcheon, 1979, op. cit. Part III, Chapter 11 (see note 9); R.T. McCutcheon, 'The role of industrialised building in Soviet Union housing policies'. *Habitat International* (forthcoming) Vol. 13(4); H.V. Morton, 'Low-cost housing in the USSR', in F. Rad Parvis *et al.* (eds), *Proceedings of the Fourth IAHS International Symposium on Housing Problems, May 1976*, New York: Pergamon Press, 1976; R.W. Davies, 'The Builders Conference 30 Nov. to 7 Dec. 1954' *Soviet Studies*, Vol. vi (4),

April 1955, pp. 444, 445 and 455.

45. McCutcheon, op. cit. Part II (see note 9).

46. See, for example, P. Trench, President NFBTE, in Civic Trust/Trades Union Congress, *Industrialized Building, Report of a Conference Held at Trade Union Congress Headquarters, Friday 29 March 1963*, Civic Trust, 1963, p. 22; White, op. cit., p. 300 (see note 21); R. Gunther, Minister of Labour, 'The Future of Industry', *Building*, Vol. 10, March 1967, p. 123.

47. 'Industrialised building—a matter of communication', *The Builder*, 13 December 1963, p. 1206.

48. Civic Trust/Trades Union Congress, op. cit., p. 23 (see note 46).

49. Report of a speech in Bexley—'Government inquiry sought', *The Builder*, 12 August 1966, p. 73.

50. McCutcheon, op. cit. Part III, Chapter 12 (see note 9). R.T. McCutcheon, 'The role of industrialised building in low income housing policy in the United States of America', *Habitat International* (forthcoming), Vol. 14(1).

51. McCutcheon, op. cit., Part IV (see note 9). During a seminar on industrialized building in Iran, one of the world's leading authorities on industrialized building—he has produced one of the textbooks on the subject—admitted, firstly, that industrialized methods could not possibly compete with the existing use of metal frame and brick infill (let alone traditional mudbrick) and, secondly, that one of the reasons why Denmark wished to provide industrialized systems to Iran was because nine-tenths of its own capacity was currently dormant. In spite of his observations he somewhat cynically conceded that he would be prepared to advise Iran on the use of industrialized systems.

52. P.F. Patman *et al.*, *Industrialized Building: A Comparative Analysis of European Experience*, Washington: Department of Housing and Urban Development (HUD), 1968, p. 3, underlined in original text.

53. Y. Samodayev, 'Housing construction in the Soviet Union', in J.S. Fuerst (ed.), *Public Housing in Europe and America*, London: Croom Helm, 1974, p. 115, Table 3.

54. B.R. Rubanenko, 'Dwellings for all', in Sfintesco (ed.), 1977, op. cit., p. 290 (see note 42).

55. Ibid., p. 287.

56. H.W. Morton, 'Low cost housing in the Soviet Union', in P.F. Rad *et al.* (eds), *Proceedings of the IAHS International Symposium on Housing Problems—1976*, Florida and Oxford: Pergamon, 1976, Vol. 1, p. 103.

57. Ibid.

58. G.F. Kuznetzov, 'Major scientific problems in the construction of prefabricated building of large sized elements', in CIB (ed.), *Building Research and Documentation: Proceedings of the 1st CIB Congress, Rotterdam, 1959*, Amsterdam: Elsevier, 1961.

59. Rubanenko, op. cit. p. 288.

60. A.J. di Maio, *Soviet Urban Housing Problems and Policies*, New York: Praeger, 1974, p. 24.

61. For greater detail see R.T. McCutcheon, 'Industrialised house-building in the U.K. 1965-1977', *Habitat International*, Vol. 13(1), 1989, pp. 33-63.

62. McCutcheon, op. cit., Part III, Chapter 11, pp. 310, 311 and 320-322 (see note 9). R.T. McCutcheon, 'The role of industrialised building in Soviet Union housing policies', *Habitat International* (forthcoming), Vol. 13(4).

63. McCutcheon, 1979, op. cit., Chapter 12 (see note 9). R.T. McCutcheon, 'The role of industrialised building in low income housing policy in the United States of America', *Habitat International* (forthcoming), Vol. 14(1).

64. McCutcheon, op. cit., Part IV, Chapter 14 (see note 9).

65. R.T. McCutcheon, 'Highrise flats in the U.K. and Iran: a comparative

review', paper presented at *Conference 2001, Urban Space for Life and Work, UNESCO, Paris, 21-25 November 1977*, Hamadan: Bin Ali Sina University, Mimeo, October 1977, 56 pp.

66. Idem, 'Industrialised house building, 1956-1976', *Habitat International*, Vol. 12(1), 1988, pp. 95-104.

67. Idem, 1979, op. cit., Part II (see note 9); see also McCutcheon, 1989, op. cit. (see note 61) and R.T. McCutcheon, 'Major participants in the U.K. (industrialized house-) building industry 1964-1977', *Habitat International*, Vol. 12 (1), 1988, pp. 105-116.

68. Idem, 1989, op. cit. (see note 61).

69. Idem, 'Technical change and social need: the case of high flats', *Research Policy*, Vol. 4, 1975, pp. 262-289.

70. Idem, 1979, op. cit., Part III, Chapter 11 (see note 9). R.T. McCutcheon, 'The role of industrialised building in Soviet Union housing policies', op. cit. (see note 62).

71. Ibid.; G. Makarevich, 'Urban planning with tall buildings in Moscow', in ASCE/IABSE, *Proceedings of the International Conference on Planning and Design of Tall Buildings*, New York: ASCE, 1973; B. Rubanenko, 'Design and construction of tall buildings', ibid., pp. 231-249. Both papers were also presented at a CIB symposium on tall buildings: *Proceedings of Symposium on Tall Buildings, CIB Report No. 21*, Moscow: Central Research and Design Institute, 1972. V. Svetlichny, *Kommunist*, No. 6, 1965; quoted by L.M. Herman, 'Urbanization and new housing construction in the USSR', in US Congress Joint Economic Committee, *Industrialized Housing Materials; Compiled and prepared for the Sub-committee on Urban Affairs of the Joint Economic Committee, Congress of the United States*, Washington: US Government Printing Office, 1969.

72. Davies, op. cit., p. 452 (see note 44); di Maio, op. cit., p. 75 (see note 60).

73. McCutcheon, op. cit., Part IV (see note 9). See also R.T. McCutcheon, 'Technology, development and environment: a case study of two resettlement schemes in Hamedan Iran' (TD/B/C.6/83), Geneva: UNCTAD, October 1982.

74. Terner and Turner, op. cit., V-i (see note 27). W. Strassman, *Building Technology and Employment in the Housing Sector of Developing Countries*, Michigan State University, 1978, p. 203. S.H.K. Yeh and A.A. Laquian (eds), *Low-Cost Housing in South East Asia: Problems, Policies and Prospects*, Ottawa: IRDC, 1978, p. 293.

75. Burns and Grebler, op. cit., p. 230 (see note 2).

76. Svetlichny, op. cit. (see note 71).

77. R.T. McCutcheon, 'Technical and economic efficiency in the UK (industrialised house-) building industry', *Habitat International*, Vol. 12(1), 1988, pp. 117-128.

78. UNECE (HOU/14), 1965, op. cit., p. 5 (see note 38).

79. UNECE (HOU/20), 1966, op. cit., pp. 164, 165 and 170 (see Ref. 25); ILO, 1968, *Social Aspects* . . ., op. cit., p. 38.

80. UNECE (ECE/364), 1959, op. cit., p. 17.

81. UNECE (HOU/14), 1965, op. cit., p. 6.

82. Ibid.

83. UNECE (HOU/36), 1969, op. cit., p. 73; underlined in text.

84. B. Lewicki, 'The economy of large panel systems', *International Journal for Housing Science and its Application*, Vol. I (1) August 1977, p. 95.

85. UNECE (HOU/14), 1965, op. cit., p. 14.

86. UNECE (HOU/20), 1966, op. cit., p. 162; underlined in text.

87. UNECE (HOU/14) 1965, op. cit., pp. 12-13; underlined in text.

88. White, op. cit., pp. 97 and 108.

89. Ibid., pp. 53, 91, 96 and 139.

90. Ibid., pp. 49–50, 154.
91. Ibid., pp. 91, 156 and 168.
92. Ibid., pp. 48 and 107.
93. UN Department of Economic and Social Affairs (UNESA), *Guidelines for Government Policies and Measures for the Gradual Industrialization of Building* (ST/ESA/7), New York: UN, 1974, p. 1.
94. Ibid., p. 39.
95. UN Centre for Housing Building and Planning and Office of Technical Cooperation, *Report of the Interregional Seminar on Design and Technology for Low-cost Housing, Budapest, Hungary, 9–20 April 1974*, New York: UN, 1975; draft (ESA/HBP/AC.13/1), p. 54.
96. Ibid.
97. Civic Trust/Trades Union Congress, 1963, op. cit., p. 22.
98. *Official Architect*, Vol. 28 (March 1965), p. 341.
99. UNECE (HOU/20), 1966, op. cit., p. 152.
100. UNECE (HOU/14), 1965, op. cit., p. 5.
101. Civic Trust/Trades Union Congress, 1963, op. cit. p. 23; see also R. Jordan Furneaux 'LCC', *Architectural Review*, Vol. 120 (November 1957), p. 503.
102. White, op. cit., p. 2.
103. Marion Bowley, *Housing and State*, London: George Allen & Unwin, 1945, p. vi.
104. White, op. cit., pp. 58, 68, 73, 123 and 174.
105. The scale of this experiment contrasts strongly with the size of research and development carried out formally by government and industry. In 1964 it was estimated that the cost of research and development amounted to 0.1 per cent of the industry's turnover. Dr Lea, the Director of the Building Research Station from 1946 to 1965, stated that by 1970 this had 'grown somewhat' but that it was still small: F. Lea, *Science and Building*, London: HMSO, 1971, p. 200.
106. *USSR Housing Construction*, Moscow. Novosti Press (no date, no pagination).
107. R.T. McCutcheon, 'Technical and economic efficiency in the U.K. (industrialised house-) building industry 1965–1977', *Habitat International*, Vol. 12(1), 1988, pp. 117–128.
108. K. Pavitt and W. Walker, 'Government policies towards industrial innovation: a review', *Research Policy*, Vol. 58, 1976, p. 42.
109. McCutcheon, 1988, op. cit.
110. Much of the discussion on self-help and appropriate technology has often concentrated upon the technological aspect of housing, in particular the shell of the house. The emphasis upon the shell not only repeats the prime error of industrialized building but ignores the extent to which adequate accommodation now includes a variety of sophisticated services. These comments apply with reduced force to developing countries. In relation to the latter, the author is more concerned that governments do not use 'self-help' as an excuse for not devoting enough resources to the housing sector.
111. R.E. Jeanes, *Study of Operative Skills: Problems, Progress and Plans; Building Research Current Papers Research Series 24*, Garston: BRS MOT, September 1964 (reported 1968), pp. 1 and 8.
112. Bowley, 1945, op. cit., p. 48.
113. Ibid., p. 41.

Notes to Table 2
Sources: *Annual Bulletin of Housing and Building Statistics for Europe,* 1967, Vol. VXI, p. 50 *et seq.*, Table 6; 1969, Vol. VXIII, p. 45 *et seq.*, Table 6; 1973, Vol. VXVII, p. 44 *et seq.*, Table 6; 1976, Vol. VXX, p. 38 *et seq.*, Table 6; and the following:

1. Monograph of Belgium in International Federation of Building and Public Works (IFBPW), *Social Aspects of Prefabrication in the Construction Industry,* IFBPW, Paris, 1967, p. 2; quoted in P.F. Patman *et al.*, *Industrialized Building: A Comparative Analysis of European Experience,* Washington: Department of Housing and Urban Development (HUD), 1968, p. 11.
2. Monograph of Denmark, IFBPW, op. cit., p. 1.
3. UN CHBP, *Industrialization of Building* (E/C.6/70/Add. 1), UN, New York, 1967, Annex 1, p. 112.
4. M. Kjeldsen, 'Introduction to Denmark' in *Iranian Danish Symposium 1976: Aims and Means of Industrialised Building,* Tehran, September 1967, Part 3; and M. Kjeldsen, *Housing in Denmark 1965-76,* Danish Building Centre, Copenhagen, 1976, p. 5.
5. A.P. Snabe, 'Large panel buildings in Denmark', in CIB (ed.), *Research into Practice: The Challenge of Application,* CIB, Paris, 1971, Vol. 1, p. 393.
6. G. Bonhomme, 'French experience in the use of heavy concrete elements', in CIB (ed.), *Building Research and Documentation,* Elsevier, Amsterdam, 1961, p. 193.
7a. V.I. Ovsyankin, 'Component building: II—large panel housing, volume in total town housing construction', in CIB (ed.), *Building Cost and Quality,* CIB, Rotterdam, 1969, p. 23.
7b. B.D. Pleissen and N.P. Rosanov, 'Large panel buildings as a percentage of total houses built', in CIB (ed.), 1971, op. cit., p. 371.
7c. Patman *et al.*, op. cit., p. 11.
8. UNECE, *Proceedings of the Seminar on Changes in the Structure of the Building Industry Necessary to Improve its Efficiency and to Increase its Output.* Prague, April 1964, (ST/ECE/HOU/13), UN, New York, 1965, Vol. 1, p. 222.
9. Monograph of Finland, IFBPW, op. cit., p. 1.
10a. Residential buildings not dwellings.
10b. Monograph of Netherlands, IFBPW, op. cit., p. 1.
11. Material used for outer wall.
12. Share of Prefabrication, Monograph of Norway, IFBPW, op. cit., p. 1.
13. Monograph of Sweden, UNECE (HOU/13), 1965, op. cit., Vol. 3, pp. 632-634.
14. Multi-dwelling houses for which State loans have been granted.
15. Ovsyankin, op. cit.
16. New construction of dwellings by local authorities only.

Notes to Table 3
Sources: UNECE, *Annual Bulletin of Housing and Building Statistics for Europe, 1967, Vol. XI, p. 50 et seq.*, Table 6; 1969, Vol. XIII, p. 45 *et seq.*, Table 6; 1973, Vol. XVII, p. 44 *et seq.*, Table 6; 1976, Vol. XX, p. 38 *et seq.*, Table 6; and the following:

1. Medium-sized panels and assembled reinforced concrete frames.
2. Concrete blocks, medium and large panels and assembled frames, UN CHBP, *Industrialization of Building* (E/C.6/70/Add. 1), UN, New York, 1967, Annex 1, p. 5.
3. Concrete blocks (up to 750 kg); concrete blocks and panels up to 2,000 kg; large panels up to 5,000 kg. UNECE, 1967, op. cit.
4. Larger residential building constructed by the State building enterprises, UNECE, 1973, op cit.
5. Industrialized housing as a per cent of urban housing; Monograph of Poland, UNECE (HOU/13), 1965, Vol. 2, p. 536.
6. E. Kuminek, 'Changes in the output of the building industry as a factor in the development of home building', in A.A. Nevitt (ed.), *The Economic Problems of Housing,* Macmillan, London, 1967, p. 232.
7. Large panel as a per cent of state financed construction, UN CHBP, *Industrialization of Building* (E/C.6/70/Add. 1), Geneva, UN, 5 Sept. 1967), Annex 1, p. 85.
8. V.I. Ovsyankin, 'Component Building: II—large panel housing', in CIB (ed.), *Building Cost and Quality,* CIB, Rotterdam, 1969.

9. Y. Samodayev, 'Housing construction in the Soviet Union', in J. Fuerst (ed.), *Public Housing in Europe and America*, Croom Helm, London, 1974, p. 115.

10. Ovsyankin, op. cit.

11. Monograph of Ukrainian SSR in UNECE (HOU/13), 1965, op. cit., Vol. 3, pp. 644–645.

12. ILO, Building Civil Engineering and Public Works Committee Eighth Session, Geneva, 1968, *Social Aspects of Prefabrication in the Construction Industry*, ILO, Geneva, 1968, p. 16.

13. Zukurov quoted in H.W. Morton, 'Low cost houses in the USSR', in F. Rad. Parvis et al. (eds.), *Proceedings of the IAHS International Symposium on Housing Problems—1976*, Pergamon, Florida and Oxford, 1976, Vol. 1, p. 105.

14. B.R. Rubanenko, 'Dwellings for all', in D. Sfintesco (ed.), *Conference 2001: Urban Space for Life and Work*, Paris: UNESCO/Council on Tall Buildings and Urban Habitat, 1977, Vol. II, p. 287.

15. G.F. Kuznetzov, 'Major scientific problems in the construction of prefabricated building of large-sized elements', in CIB (ed.), *Building Research and Documentation*, Elsevier, Amsterdam, 1965.

16. Material used for outer wall.

17. UN, *Industrialization of Building*, 1967, op. cit., Annex 1, pp. 101 and 104.

18. G. Blachere, 'Analytical report', in *Proceedings of the Seminar on Changes in the Structure of the Building Industry Necessary to Improve Its Efficiency and to Increase Its Output, Prague, Czechoslovakia, 19-30 April 1964* (ST/ECE/HOU/13), UN, New York, 1965, Vol. 1, p. 223.

19. L.M. Herman, 'Urbanization and new housing construction in the USSR', in US Congress Joint Economic Committee, *Industrialized Housing, Materials Compiled and Prepared for the Subcommittee on Urban Affairs of the Joint Economic Committee, Congress of the United States*, US Government Printing Office, Washington, DC, 1969, p. 28.

Book Review

André Guillerme, *Le Temps de l'Eau: La Cité, L'Eau et les Techniques: Nord de la France Fin IIIe-Début XIXe Siècle*, Seyssel, 1983. ISBN: 2-903528-22-5.
Translated into English as *The Age of Water: The Urban Environment in the North of France, AD 300-1800*, College Station, 1988. ISBN: 0-89096-270-7

Unless one lives in Norway or in Scotland with their hydroelectric power stations or has the historical imagination to see the significance of the somewhat twee watermills now displayed in 'heritage' centres, then one's perception of water is radically different from that of the Middle Ages. While this book covers the use of water in northern France from the end of the Roman Empire until the end of the first French Empire, it is particularly instructive to compare, using this book, the uses of water in the Middle Ages with those today.

Everyone recognizes that fresh water and, symbiotically as important, the disposal of waste products, are central to our physical well-being, hence the recent controversies over the demunicipalization of the water companies in Britain. But until the advent of the steam engine (itself using water, let us not forget), and subsequently of the use of gas and electricity, the vast bulk of motive mechanical power from about AD 300 was derived directly from water. Additionally, as Guillerme shows in this structuralist study, water had significance even beyond this. For the defence of towns and for its role in developing town structures water played a critical and in many cases a crucial role. Nowadays water no longer forms any sort of effective barrier (merely, in the form of the English Channel, a nuisance to trade and travel soon to be eliminated) and its role in town planning is now minimal (witness the proposed road tunnel to be built along the bed of the Thames). For internal trade, at least in England, water plays no significant role and only in international trade (but not travel) does water continue to play a major role. Water now has uses which would have been inconceivable during the Middle Ages. For example, the use of water for leisure purposes (from sailing to living in expensive waterside properties) is a fairly recent innovation. Only perhaps in drinking and in religion (and possibly washing) would those in the Middle Ages recognize any similarities in the use and significance of water then and now. The perception of water has changed out of all recognition during the past millennium.

For a period during the Middle Ages water was one of the chief agents of social and technological change. Thus water was one of the contributing factors that led to the full-scale industrialization of Western European society, which has now come to affect, by one means or another, every individual on earth.

It is these themes that are central to Guillerme's book, whose English translation is especially welcome. Because it is a structuralist study it is not obsessed, as much history of technology is, with origins, but is concerned with how water, in all its uses, affected the daily lives of the inhabitants of northern France.

The opening chapter on the sacred uses of water in late antiquity sets the scene for the importance of water. Guillerme shows how the local river gods of the Romans had various Christian guises imposed on them, a process that led to the founding of festivals such as Rogation days. Curiously, following this the role of the Church and, indeed, Christianity receives little discussion. Guillerme prefers to concentrate on secular causes (e.g. war) to account for the modification of water technology. Yet he quotes figures (page 95 of the English translation) which demonstrate that Church institutions controlled over half the water mills in northern France during the Middle Ages. These figures he says confirm the importance of 'both lay and ecclesiastical feudalism in the matter of technical innovations', and only on the final page does he refer to the role of the Benedictines in developing water power.

This is a fascinating area of study (as has been shown by the late Lynn White) and it would have benefited Guillerme's book greatly if he had discussed how it was that monks, devoted to the Divine Office, came to be among the leading technical innovators of the Middle Ages. It would have not only provided him with an additional organizing principle for the book, but also it would have been worth exploring to see how what we now take to be the prototype of the Puritan work ethic affected technical practice.

One religious group that Guillerme does discuss, and rightly so, is the Jewish communities in the towns he analyses. Guillerme argues that their hygienic practices greatly influenced the practices of the whole community during the Middle Ages. Furthermore, he suggests that increasing attention to individual hygiene during the Middle Ages, particularly in washing, was also a consequence of Jewish practices.

This book is well and profusely illustrated, particularly with maps of the northern French towns showing the development of their watercourses over time. But what does all this, and much more contained in this book, tell us about society and technical innovation in the Middle Ages? Guillerme demonstrates quite conclusively that new problems for cities that emerged during this period (e.g. sieges) were, at least partially, met with water orientated solutions (i.e. moats). He also demonstrates that water was central to people's experience in many more aspects of their lives than it is today, and that great importance was placed on the controls society put on the use of water.

But to what extent can we say that water was a cause of this change and to what extent was it an agent of change being used to serve other interests? Is it right, in other words, to call this period the 'Age of Water'? From the point of view of technology this is indisputably so, but from the point of view of history (for we must be historians first, before considering our individual specialities) this is less certain. The Church, both materially and spiritually, dominated the use of water. No doubt it did this for its own ends, but on the whole Guillerme leaves us in darkness about what these ends were.

To end on a more positive note, the structuralist nature of this study makes for a refreshing approach to technology. It brings forward evidence that might have been previously ignored or not looked for. More importantly it is a work that can be integrated into more general studies to give a more balanced view of the Middle Ages.

Frank A.J.L. James

Contents of Former Volumes

FIRST ANNUAL VOLUME, 1976

D.S.L. CARDWELL and RICHARD L. HILLS, Thermodynamics and Practical Engineering in the Nineteenth Century.

JACQUES HEYMAN, Couplet's Engineering Memoirs, 1726–33.

NORMAN A.F. SMITH, Attitudes to Roman Engineering and the Question of the Inverted Siphon.

R.A. BUCHANAN, The Promethean Revolution: Science, Technology and History.

M. DAUMAS, The History of Technology: its Aims, its Limits, its Methods.

KEITH DAWSON, Electromagnetic Telegraphy: Early Ideas, Proposals and Apparatus.

MARIE BOAS HALL, The Strange Case of Aluminium.

G. HOLLISTER-SHORT, Leads and Lags in Late Seventeenth-century English Technology.

SECOND ANNUAL VOLUME, 1977

EMORY L. KEMP, Samuel Brown: Britain's Pioneer Suspension Bridge Builder.

DONALD R. HILL, The Banū Mūsà and their 'Book of Ingenious Devices'.

J.F. CAVE, A Note on Roman Metal Turning.

J.A. GARCIA-DIEGO, Old Dams in Extremadura.

G. HOLLISTER-SHORT, The Vocabulary of Technology.

RICHARD L. HILLS, Museums, History and Working Machines.

DENIS SMITH, The Use of Models in Nineteenth-century British Suspension Bridge Design.

NORMAN A.F. SMITH, The Origins of the Water Turbine and the Invention of its Name.

THIRD ANNUAL VOLUME, 1978

JACK SIMMONS, Technology in History.

R.A. BUCHANAN, History of Technolgy in the Teaching of History.

P.B. MORICE, The Role of History in a Civil Engineering Course.

JOYCE BROWN, Sir Proby Cautley (1802–71), a Pioneer of Indian Irrigation.

A. RUPERT HALL, On knowing, and knowing how to . . .

FRANK D. PRAGER, Vitruvius and the Elevated Aqueducts.

JAMES A. RUFFNER, Two Problems in Fuel Technology.

JOHN C. SCOTT, The Historical Development of Theories of Wave-Calming using Oil.

FOURTH ANNUAL VOLUME, 1979

P.S. BARDELL, Some Aspects of the History of Journal Bearings and their Lubrication.

K.R. FAIRCLOUGH, The Waltham Pound Lock.

ROBERT FRIEDEL, Parkesine and Celluloid: The Failure and Success of the First Modern Plastic.

J.G. JAMES, Iron Arched Bridge Designs in Pre-Revolutionary France.

L.J. JONES, The Early History of Mechanical Harvesting.

G. HOLLISTER-SHORT, The Sector and Chain: An Historical Enquiry.

FIFTH ANNUAL VOLUME, 1980

THOMAS P. HUGHES, The Order of the Technological World.

THORKILD SCHIØLER, Bronze Roman Pistol Pumps.

STILLMAN DRAKE, Measurement in Galileo's Science.

L.J. JONES, John Ridley and the South Australian 'Stripper'.

D.G. TUCKER, Emile Lamm's Self-Propelled Tramcars 1870–72 and the Evolution of the Fireless Locomotive.

S.R. BROADBRIDGE, British Industry in 1767: Extracts from a Travel Journal of Joseph Banks.

RICHARD L. HILLS, Water, Stampers and Paper in the Auvergne: A Medieval Tradition.

SIXTH ANNUAL VOLUME, 1981

MARJORIE NICE BOYER, Moving Ahead with the Fifteenth Century: New Ideas in Bridge Construction at Orleans.

ANDRÉ WEGENER SLEESWYK, Hand-Cranking in Egyptian Antiquity.

CHARLES SÜSSKIND, The Invention of Computed Tomography.

RICHARD L. HILLS, Early Locomotive Building near Manchester.

L.L. COATSWORTH, B.I. KRONBERG and M.C. USSELMAN, The Artefact as Historical Document. Part 1: The Fine Platinum Wires of W.H. Wollaston.

A. RUPERT HALL and N.C. RUSSELL, What about the Fulling-Mill?

MICHAEL FORES, *Technik:* Or Mumford Reconsidered.

SEVENTH ANNUAL VOLUME, 1982

MARJORIE NICE BOYER, Water Mills: a Problem for the Bridges and Boats of Medieval France.

Wm. DAVID COMPTON, Internal-combustion Engines and their Fuel: a Preliminary Exploration of Technological Interplay.

F.T. EVANS, Wood Since the Industrial Revolution: a Strategic Retreat?

MICHAEL FORES, Francis Bacon and the Myth of Industrial Science.

D.G. TUCKER, The Purpose and Principles of Research in an Electrical Manufacturing Business of Moderate Size, as Stated by J.A. Crabtree in 1930.

ROMAN MALINOWSKI, Ancient Mortars and Concretes: Aspects of their Durability.

V. FOLEY, W. SOEDEL, J. TURNER and B. WILHOITE, The Origin of Gearing.

EIGHTH ANNUAL VOLUME, 1983

W. ADDIS, A New Approach to the History of Structural Engineering.

HANS-JOACHIM BRAUN, The National Association of German-American Technologists and Technology Transfer between Germany and the United States, 1884–1930.

W. BERNARD CARLSON, Edison in the Mountains: the Magnetic Ore Separation Venture, 1879–1900.

THOMAS DAY, Samuel Brown: His Influence on the Design of Suspension Bridges.

ROBERT H.J. SELLIN, The Large Roman Water Mill at Barbegal (France).

G. HOLLISTER-SHORT, The Use of Gunpowder in Mining: A Document of 1627.

MIKULÁŠ TEICH, Fermentation Theory and Practice: the Beginnings of Pure Yeast Cultivation and English Brewing, 1883–1913.

GEORGE TIMMONS, Education and Technology in the Industrial Revolution.

NINTH ANNUAL VOLUME, 1984

P.S. BARDELL, The Origins of Alloy Steels.

MARJORIE NICE BOYER, A Fourteenth-Century Pile Driver: the *Engin* of the Bridge at Orleans.

MICHAEL DUFFY, Rail Stresses, Impact Loading and Steam Locomotive Design.

JOSÉ A. GARCIA-DIEGO, Giovanni Francesco Sitoni, an Hydraulic Engineer of the Renaissance.

DONALD R. HILL, Information on Engineering in the Works of Muslim Geographers.

ROBERT J. SPAIN, The Second-Century Romano-British Watermill at Ickham, Kent.

IAN R. WINSHIP, The Gas Engine in British Agriculture, c. 1870–1925.

TENTH ANNUAL VOLUME, 1985

D. de COGAN, Dr E.O.W. Whitehouse and the 1858 trans-Atlantic Cable.

A. RUPERT HALL, Isaac Newton's Steamer.

G.J. HOLLISTER-SHORT, Gunpowder and Mining in Sixteenth- and Seventeenth-Century Europe.

C.J. JACKSON, Evidence of American Influence on the Designs of Nineteenth-Century Drilling Tools, Obtained from British Patent Specifications and Other Sources.

JACQUES PAYEN, Beau de Rochas Devant la Technique et l'Industrie de son Temps.

ÖRJAN WIKANDER, Archaeological Evidence for Early Water-Mills—an Interim Report.

A.P. WOOLRICH, John Farey and the Smeaton Manuscripts.

MIKE CHRIMES, Bridges: a Bibliography of Articles Published in Scientific Periodicals 1800–1829.

ELEVENTH ANNUAL VOLUME, 1986

HANS-JOACHIM BRAUN, Technology Transfer Under Conditions of War: German Aero-technology in Japan During the Second World War.

VERNARD FOLEY, with SUSAN CANGANELLI, JOHN CONNOR and DAVID RADER, Using the Early Slide-rest.

J.G. JAMES, The Origins and Worldwide Spread of Warren-truss Bridges in the Mid-nineteenth Century. Part 1: Origins and Early Examples in the UK.

ANDREW NAHUM, The Rotary Aero Engine.

DALE H. PORTER, An Historian's Judgments About the Thames Embankment.

JOHN H. WHITE, More Than an Idea Whose Time Has Come: The Beginnings of Steel Freight Cars.

IAN R. WINSHIP, The Acceptance of Continuous Brakes on Railways in Britain.